Mechanics of Sheet Metal Forming

Mechanics of Sheet Metal Forming

Z. Marciniak
The Technical University of Warsaw, Poland

J.L. Duncan
The University of Auckland, New Zealand

S.J. Hu
The University of Michigan, USA

OXFORD AMSTERDAM BOSTON LONDON NEW YORK PARIS
SAN DIEGO SAN FRANCISCO SINGAPORE SYDNEY TOKYO

Butterworth-Heinemann
An imprint of Elsevier Science
Linacre House, Jordan Hill, Oxford OX2 8DP
225 Wildwood Avenue, Woburn, MA 01801-2041

First published by Edward Arnold, London, 1992
Second edition published by Butterworth-Heinemann 2002

British Library Cataloguing in Publication Data
A catalogue record for this book is available from the British Library

Library of Congress Cataloguing in Publication Data
A catalogue record for this book is available from the Library of Congress

ISBN 0 7506 5300 0

Transferred to digital printing 2005

PLANT A TREE
FOR EVERY TITLE THAT WE PUBLISH, BUTTERWORTH-HEINEMANN WILL PAY FOR BTCV TO PLANT AND CARE FOR A TREE.

Contents

Preface to the second edition

The first edition of this book was published a decade ago; the Preface stated the objective in the following way.

> In this book, the theory of engineering plasticity is applied to the elements of common sheet metal forming processes. Bending, stretching and drawing of simple shapes are analysed, as are certain processes for forming thin-walled tubing. Where possible, the limits governing each process are identified and this entails a detailed study of tensile instability in thin sheet.
> To the authors' knowledge, this is the first text in English to gather together the mechanics of sheet forming in this manner. It does, however, draw on the earlier work of, for example, Swift, Sachs, Fukui, Johnson, Mellor and Backofen although it is not intended as a research monograph nor does it indicate the sources of the models. It is intended for the student and the practitioner although it is hoped that it will also be of interest to the researcher.

This second edition keeps to the original aim, but the book has been entirely rewritten to accommodate changes in the field and to overcome some earlier deficiencies. Professor Hu joined the authors and assisted in this revision. Worked examples and new problems (with sample solutions) have been added as well as new sections including one on hydroforming. Some of the original topics have been omitted or given in an abbreviated form in appendices.

In recent years, enormous progress has been made in the analysis of forming of complex shapes using finite element methods; many engineers are now using these systems to analyse forming of intricate sheet metal parts. There is, however, a wide gulf between the statement of the basic laws governing deformation in sheet metal and the application of large modelling packages. This book is aimed directly at this middle ground. At the one end, it assumes a knowledge of statics, stress, strain and models of elastic deformation as contained in the usual strength of materials courses in an engineering degree program. At the other end, it stops short of finite element analysis and develops what may be called 'mechanics models' of the basic sheet forming operations. These models are in many respects similar to the familiar strength of materials models for bending, torsion etc., in that they are applied to simple shapes, are approximate and often contain simplifying assumptions that have been shown by experience to be reasonable. This approach has proved helpful to engineers entering the sheet metal field. They are confronted with an industry that appears to be based entirely on rules and practical experience and they require some assistance to see how their engineering training can be applied to the design of tooling and to the solution of problems in the stamping plant. Experienced sheet metal engineers also find the approach useful in conceptual design, in making quick calculations in the course of more extensive design work, and in interpreting and understanding the finite element simulation results. Nevertheless, users of these models should be aware of the assumptions and limitations of these approximate models as real sheet metal designs can be much more complex than what is captured by the models.

The order in which topics are presented has been revised. It now follows a pattern developed by the authors for courses given at graduate level in the universities and to sheet metal engineers and mechanical metallurgists in industry and particularly in the automotive field. The aim is to bring students as quickly as possible to the point where they can analyse simple cases of common processes such as the forming of a section in a typical stamping. To assist in tutorial work in these courses, worked examples are given in the text as well as exercises at the end of each chapter. Detailed solutions of the exercises are given at the end of the text. The possibility of setting interesting problems is greatly increased by the familiarity of students with computer tools such as spread sheets. Although not part of this book, it is possible to go further and develop animated models of processes such as bending, drawing and stamping in which students can investigate the effect of changing variables such as friction or material properties. At least one package of this kind is available through Professor Duncan and Professor Hu.

Many students and colleagues have assisted the authors in this effort to develop a sound and uncomplicated base for education and the application of engineering in sheet metal forming. It is impossible to list all of these, but it is hoped that they will be aware of the authors' appreciation. The authors do, however, express particular thanks to several who have given invaluable help and advice, namely, A.G. Atkins, W.F. Hosford, F. Wang, J. Camelio and the late R. Sowerby. In addition, others have provided comment and encouragement in the final preparation of the manuscript, particularly M. Dingle and R. Andersson; the authors thank them and also the editorial staff at Butterworth-Heinemann.

J.L. Duncan
Auckland

S.J. Hu
Ann Arbor
2002

Preface to the first edition

In this book, the theory of engineering plasticity is applied to the elements of common sheet metal forming processes. Bending, stretching and drawing of simple shapes are analysed, as are certain processes for forming thin-walled tubing. Where possible, the limits governing each process are identified and this entails a detailed study of tensile instability in thin sheet.

To the author's knowledge, this is the first text in English to gather together the mechanics of sheet forming in this manner. It does, however, draw on the earlier work of, for example, Swift, Sachs Fukui, Johnson, Mellor and Backofen although it is not intended as a research monograph nor does it indicate the sources of the models. It is intended for the student and the practitioner although it is hoped that it will also be of interest to the researcher.

In the first two chapters, the flow theory of plasticity and the analysis of proportional large strain processes are introduced. It is assumed that the reader is familiar with stress and strain and the mathematical manipulations presented in standard texts on the basic mechanics of solids. These chapters are followed by a detailed study of tensile instability following the Marciniak–Kuczynski theory. The deformation in large and small radius bends is studied and an approximate but useful approach to the analysis of axisymmetric shells is introduced and applied to a variety of stretching and drawing processes. Finally, simple tube drawing processes are analysed along with energy methods used in some models.

A number of exercises are presented at the end of the book and while the book is aimed at the engineer in the sheet metal industry (which is a large industry encompassing automotive, appliance and aircraft manufacture) it is also suitable as a teaching text and has evolved from courses presented in many countries.

Very many people have helped with the book and it is not possible to acknowledge each by name but their contributions are nevertheless greatly appreciated. One author (J.L.D.) would like to thank especially his teacher W. Johnson, his good friend and guide over many years R. Sowerby, the illustrator S. Stephenson and, by no means least, Mrs Joy Wallace who typed the final manuscript.

Z. Marciniak, Warsaw
J.L. Duncan, Auckland
1991

Disclaimer

The purpose of this book is to assist students in understanding the mechanics of sheet metal forming processes. Many of the relationships are of an approximate nature and may be unsuitable for engineering design calculations. While reasonable care has been taken, it is possible that errors exist in the material contained and neither the authors nor the publisher can accept responsibility for any results arising from use of information in this book.

Introduction

Modern continuous rolling mills produce large quantities of thin sheet metal at low cost. A substantial fraction of all metals are produced as thin hot-rolled strip or cold-rolled sheet; this is then formed in secondary processes into automobiles, domestic appliances, building products, aircraft, food and drink cans and a host of other familiar products. Sheet metals parts have the advantage that the material has a high elastic modulus and high yield strength so that the parts produced can be stiff and have a good strength-to-weight ratio.

A large number of techniques are used to make sheet metal parts. This book is concerned mainly with the basic mechanics that underlie all of these methods, rather than with a detailed description of the overall processes, but it is useful at this stage to review briefly the most common sheet forming techniques.

Common forming processes

Blanking and piercing. As sheet is usually delivered in large coils, the first operation is to cut the blanks that will be fed into the presses; subsequently there may be further blanking to trim off excess material and pierce holes. The basic cutting process is shown in Figure I.1. When examined in detail, it is seen that blanking is a complicated process of plastic shearing and fracture and that the material at the edge is likely to become hardened locally. These effects may cause difficulty in subsequent operations and information on tooling design to reduce problems can be found in the appropriate texts.

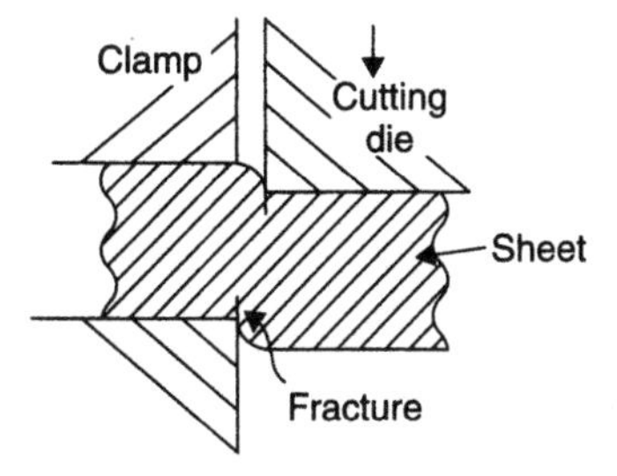

Figure I.1 Magnified section of blanking a sheet showing plastic deformation and cracking.

Bending. The simplest forming process is making a straight line bend as shown in Figure I.2. Plastic deformation occurs only in the bend region and the material away from the bend is not deformed. If the material lacks ductility, cracking may appear on

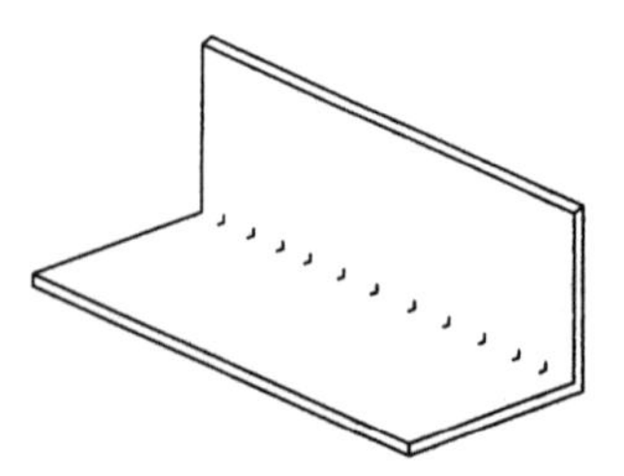

Figure I.2 Straight line bend in a sheet.

the outside bend surface, but the greatest difficulty is usually to obtain an accurate and repeatable bend angle. Elastic springback is appreciable.

Various ways of bending along a straight line are shown in Figure I.3. In *folding* (a), the part is held stationary on the left-hand side and the edge is gripped between movable tools that rotate. In *press-brake* forming (b), a punch moves down and forces the sheet into a vee-die. Bends can be formed continuously in long strip by *roll forming* (c). In roll forming machines, there are a number of sets of rolls that incrementally bend the sheet, and wide panels such as roofing sheet or complicated channel sections can be made in this process. A technique for bending at the edge of a stamped part is *flanging* or *wiping* as shown in Figure I.3(d). The part is clamped on the left-hand side and the flanging tool moves downwards to form the bend. Similar tooling is used is successive processes to bend the sheet back on itself to form a *hem*.

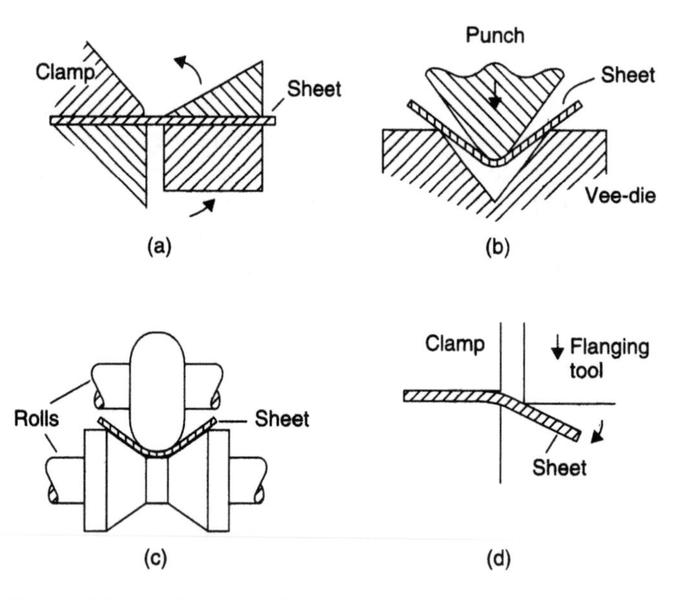

Figure I.3 (a) Bending a sheet in a folding machine. (b) Press brake bending in a vee-die. (c) Section of a set of rolls in a roll former. (d) Wiping down a flange.

If the bend is not along a straight line, or the sheet is not flat, plastic deformation occurs not only at the bend, but also in the adjoining sheet. Figure I.4 gives examples. In *shrink flanging* (a), the edge is shortened and the flange may buckle. In *stretch flanging* (b), the length of the edge must increase and splitting could be a problem. If the part is curved near the flange or if both the flange and the part are curved, as in Figure I.4(c), the flange may be either stretched or compressed and some geometric analysis is needed to determine this. All these flanges are usually formed with the kind of tooling shown in Figure I.3(d).

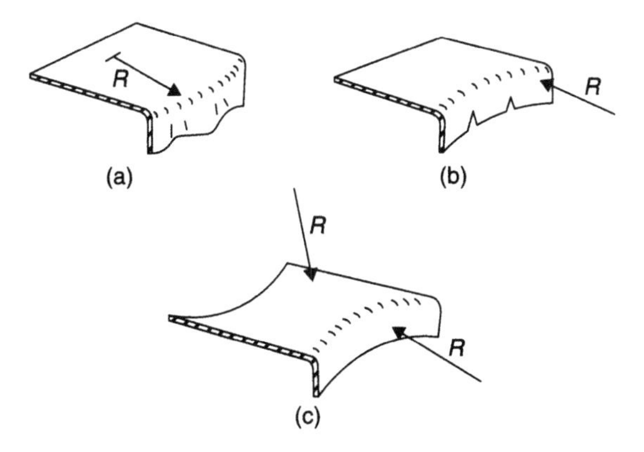

Figure I.4 (a) A shrink flange showing possible buckling. (b) A stretch flange with edge cracking. (c) Flanging a curved sheet.

Section bending. In Figure I.5, a more complicated shape is bent. At the left-hand end of the part, the flange of the channel is stretched and may split, and the height of the leg, h, will decrease. When the flange is on the inside, as on the right, wrinkling is possible and the flange height will increase.

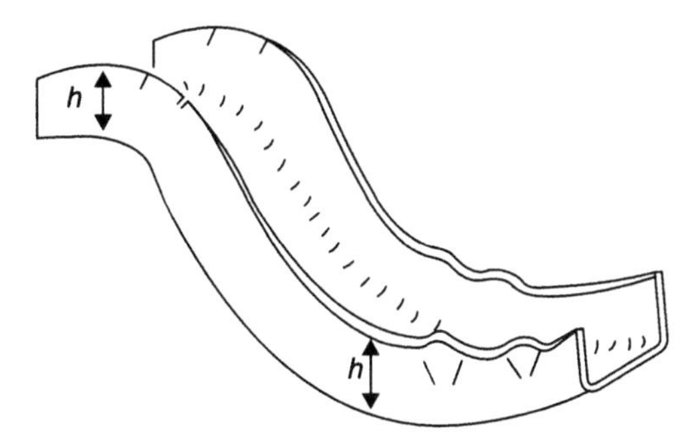

Figure I.5 Inside and outside bends in a channel section.

Stretching. The simplest stretching process is shown in Figure I.6. As the punch is pushed into the sheet, tensile forces are generated at the centre. These are the forces that cause the deformation and the contact stress between the punch and the sheet is very much lower than the yield stress of the sheet.

The tensile forces are resisted by the material at the edge of the sheet and compressive hoop stresses will develop in this region. As there will be a tendency for the outer region to buckle, it will be held by a blank-holder as shown in Figure I.6(b). The features mentioned are common in many sheet processes, namely that forming is not caused by the direct

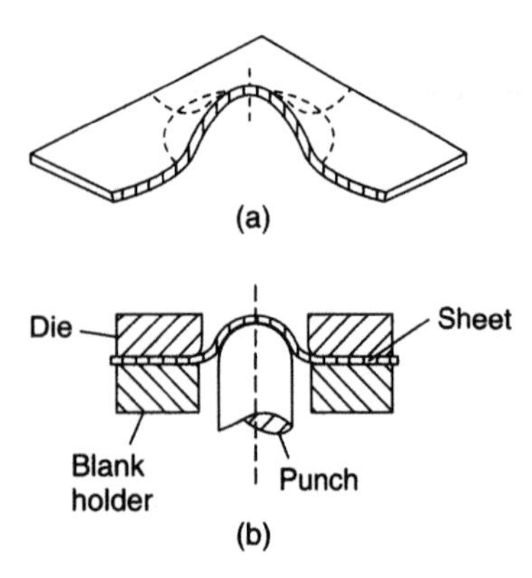

Figure I.6 (a) Stretching a dome in a sheet. (b) A domed punch and die set for stretching a sheet.

contact stresses, but by forces transmitted through the sheet and there will be a balance between tensile forces over the punch and compressive forces in the outer flange material.

Hole extrusion. If a hole smaller than the punch diameter is first pierced in the sheet, the punch can be pushed through the sheet to raise a lip as in the hole extrusion in Figure I.7. It will be appreciated that the edge of the hole will be stretched and splitting will limit the height of the extrusion.

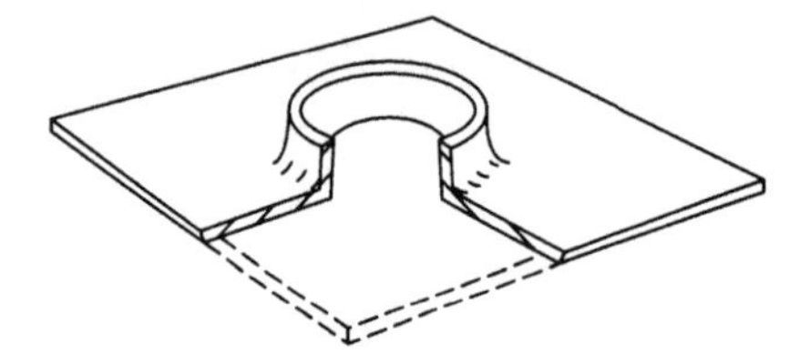

Figure I.7 Extrusion of a punched hole using tooling similar to Figure I.6(b).

Stamping or draw die forming. The part shown in Figure I.8(a) is formed by stretching over a punch of more complicated shape in a *draw die*. This consists of a *punch*, and *draw ring* and *blank-holder* assembly, or *binder*. The principle is similar to punch stretching described above, but the outer edge or flange is allowed to draw inwards under restraint to supply material for the part shape. This process is widely used to form auto-body panels and a variety of appliance parts. Much of the outer flange is trimmed off after forming so that it is not a highly efficient process, but with well-designed tooling, vast quantities of parts can be made quickly and with good dimensional control. Die design requires the combination of skill and extensive computer-aided engineering systems, but for the purpose of conceptual design and problem solving, the complicated deformation system can be broken down into basic elements that are readily analysed. In this book, the analysis of these macroscopic elements is studied and explained, so that the reader can understand those factors that govern the overall process.

Deep drawing. In stamping, most of the final part is formed by stretching over the punch although some material around the sides may have been drawn inwards from the flange. As

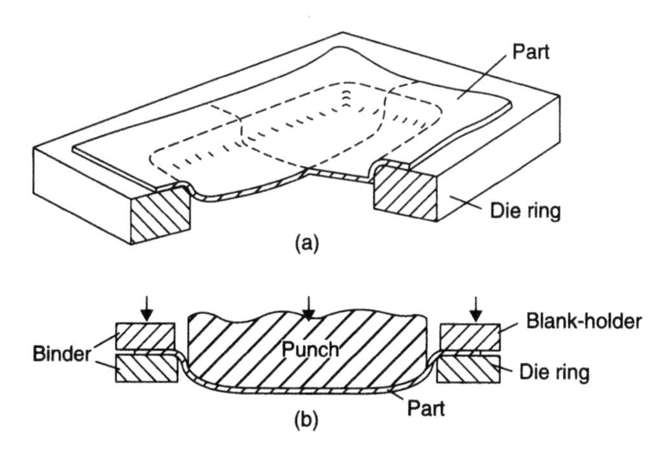

Figure I.8 (a) Typical part formed in a stamping or draw die showing the die ring, but not the punch or blankholder. (b) Section of tooling in a draw die showing the punch and binder assembly.

there is a limit to the stretching that is possible before tearing, stamped parts are typically shallow. To form deeper parts, much more material must be drawn inwards to form the sides and such a process is termed *deep drawing*. Forming a simple cylindrical cup is shown in Figure I.9. To prevent the flange from buckling, a blankholder is used and the clamping force will be of the same order as the punch force. Lubrication is important as the sheet must slide between the die and the blankholder. Stretching over the punch is small and most of the deformation is in the flange; as this occurs under compressive stresses, large strains are possible and it is possible to draw a cup whose height is equal to or possibly a little larger than the cup diameter. Deeper cups can be made by *redrawing* as shown in Figure I.10.

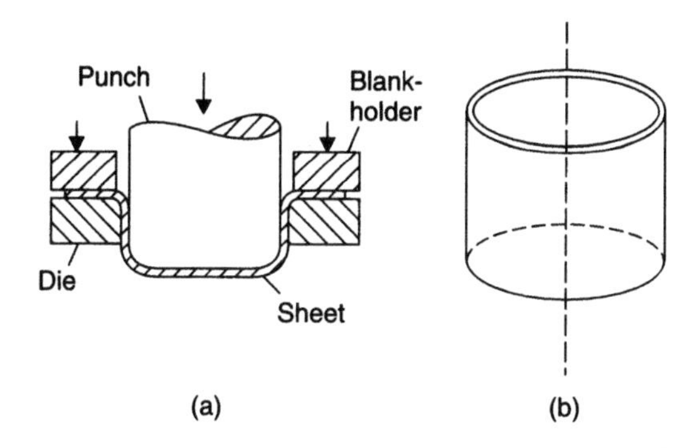

Figure I.9 (a) Tooling for deep drawing a cylindrical cup. (b) Typical cup deep drawn in a single stage.

Tube forming. There are a number of processes for forming tubes such as *flaring* and *sinking* as shown in Figure I.11. Again, these operations can be broken down into a few elements, and analysed as steady-state processes.

Fluid forming. Some parts can be formed by fluid pressure rather than by rigid tools. Quite high fluid pressures are required to form sheet metal parts so that equipment can

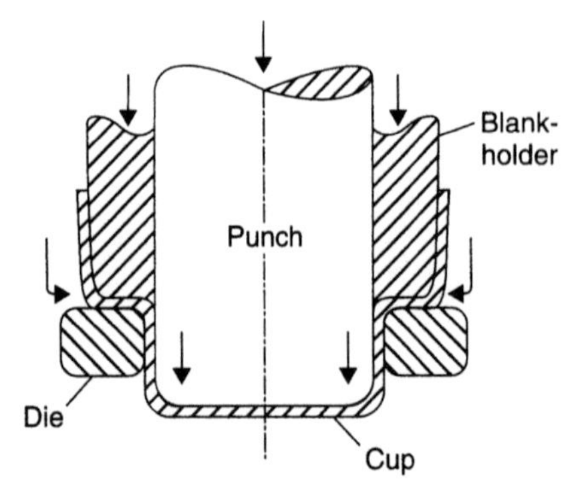

Figure I.10 Section of tooling for forward redrawing of a cylindrical cup.

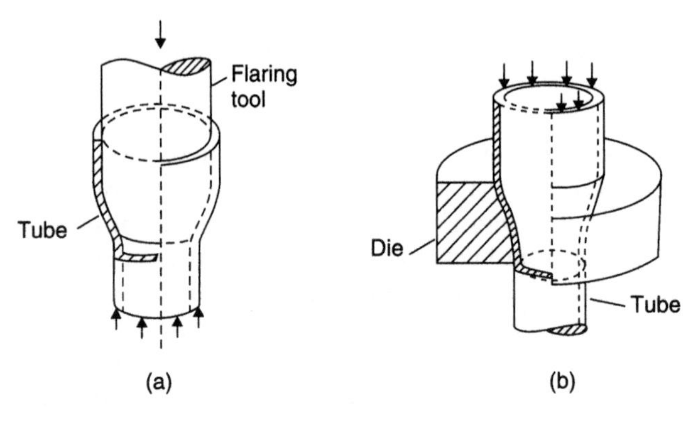

Figure I.11 (a) Expanding the end of a tube with a flaring tool. (b) Reducing the diameter of a tube by pressing it through a sinking die.

be expensive, but savings in tooling costs are possible and the technique is suitable where limited numbers of parts are required. For forming flat parts, a diaphragm is usually placed over the sheet and pressurized in a container as in Figure I.12. As the pressure to form the sheet into sharp corners can be very high, the forces needed to keep the container closed are much greater than those acting on a punch in a draw die, and special presses are required. Complicated tubular parts for plumbing fittings and bicycle frame brackets are made by a combination of fluid pressure and axial force as in Figure I.13. Tubular parts, for example frame structures for larger vehicles, are made by bending a circular tube, placing it in a closed die and forming it to a square section as illustrated in Figure I.14.

Coining and ironing. In all of the processes above, the contact stress between the sheet and the tooling is small and, as mentioned, deformation results from membrane forces in the sheet. In a few instances, through-thickness compression is the principal deformation force. *Coining*, Figure I.15, is a local forging operation used, for example, to produce a groove in the lid of a beverage can or to thin a small area of sheet. *Ironing*, Figure I.16, is a continuous process and often accompanies deep drawing. The cylindrical cup is forced through an ironing die that is slightly smaller than the punch plus metal thickness dimension. Using several dies in tandem, the wall thickness can be reduced by more than one-half in a single stroke.

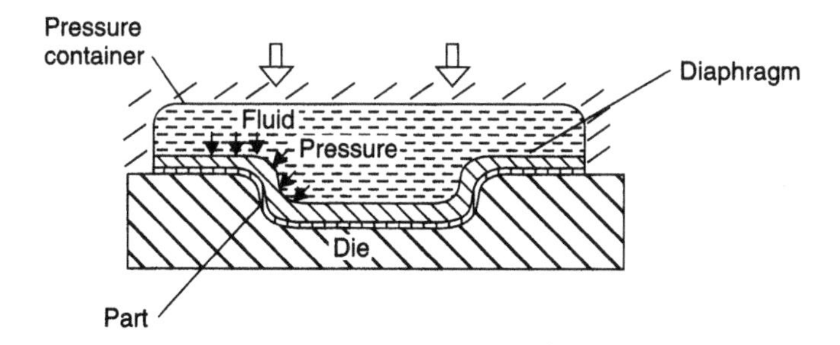

Figure I.12 Using fluid pressure (hydroforming) to form a shallow part.

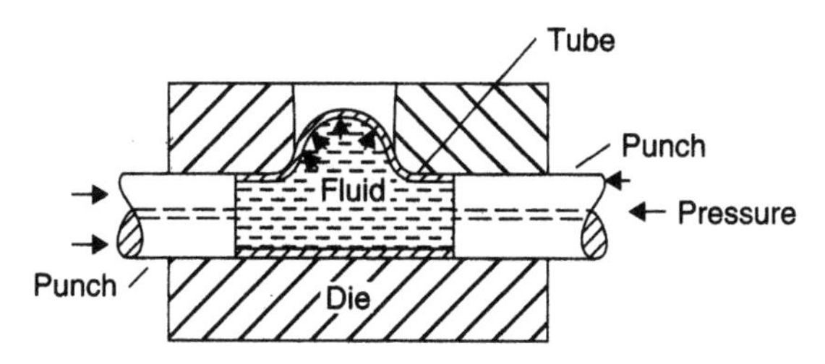

Figure I.13 Using combined axial force and fluid pressure to form a plumbing fitting (tee joint).

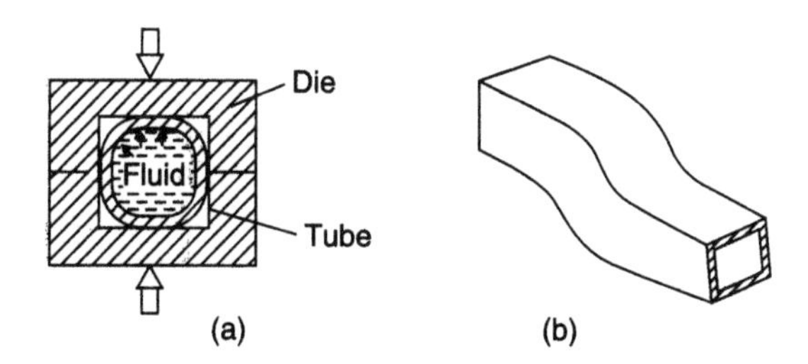

Figure I.14 (a) Expanding a round tube to a square section in a high pressure hydroforming process. (b) A section of a typical hydroformed part in which a circular tube was pre-bent and then formed by fluid pressure in a die to a square section.

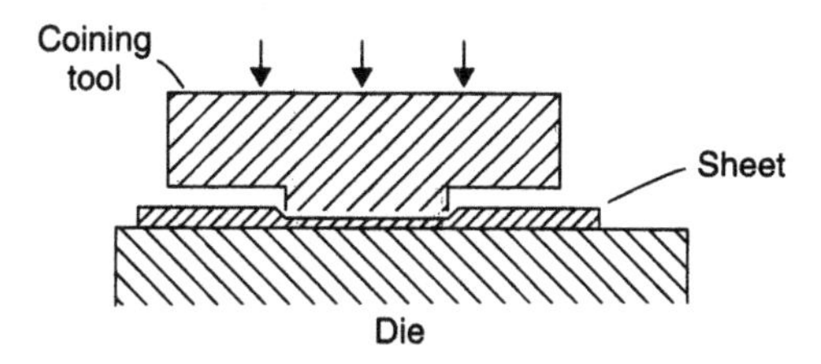

Figure I.15 Thinning a sheet locally using a coining tool.

Summary. Only very simple examples of industrial sheet forming processes have been shown here. An industrial plant will contain many variants of these techniques and numerous presses and machines of great complexity. It would be an overwhelming task to deal with all the details of tool and process design, but fortunately these processes are all made up of relatively few elemental operations such as stretching, drawing, bending, bending

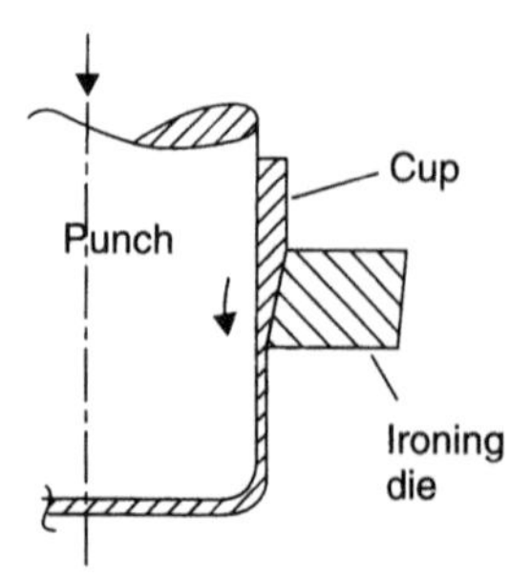

Figure I.16 Thinning the wall of a cylindrical cup by passing it through an ironing die.

under tension and sliding over a tool surface. Each of the basic deformation processes can be analysed and described by a 'mechanics model', i.e. a model similar to the familiar ones in elastic deformation for tension in a bar, bending of a beam or torsion of a shaft; these models form the basis for mechanical design in the elastic regime. This book presents similar models for the deformation of sheet. In this way, the engineer can apply a familiar approach to problem solving in sheet metal engineering.

Application to design

The objective in studying the basic mechanics of sheet metal forming is to apply this to part and tool design and the diagnosis of plant problems. It is important to appreciate that analysis is only one part of the design process. The first step in design is always to determine *what* is required of the part or process, i.e. its function. Determining *how* to achieve this comes later. When the function is described completely and in quantitative terms, the designer can then address the 'how'. This is typically an iterative process in which the designer makes some decision and then determines the consequences. A good designer will have a feeling for the consequences before any calculations are made and this ability is derived from an understanding of the basic principles governing each operation. Once the decision is made, simple and approximate calculations are usually sufficient to justify the decision. There will be a point when an extensive and detailed analysis is needed to confirm and prove the design, but this book is aimed at the initial but important stage of the process, namely being able to understand the mechanics of sheet forming processes and then analysing these in a quick and approximate manner.

1 Material properties

The most important criteria in selecting a material are related to the function of the part – qualities such as strength, density, stiffness and corrosion resistance. For sheet material, the ability to be shaped in a given process, often called its *formability*, should also be considered. To assess formability, we must be able to describe the behaviour of the sheet in a precise way and express properties in a mathematical form; we also need to know how properties can be derived from mechanical tests. As far as possible, each property should be expressed in a fundamental form that is independent of the test used to measure it. The information can then be used in a more general way in the models of various metal forming processes that are introduced in subsequent chapters.

In sheet metal forming, there are two regimes of interest – elastic and plastic deformation. Forming a sheet to some shape obviously involves permanent 'plastic' flow and the strains in the sheet could be quite large. Whenever there is a stress on a sheet element, there will also be some elastic strain. This will be small, typically less than one part in one thousand. It is often neglected, but it can have an important effect, for example when a panel is removed from a die and the forming forces are unloaded giving rise to elastic shape changes, or 'springback'.

1.1 Tensile test

For historical reasons and because the test is easy to perform, many familiar material properties are based on measurements made in the tensile test. Some are specific to the test and cannot be used mathematically in the study of forming processes, while others are fundamental properties of more general application. As many of the specific, or non-fundamental tensile test properties are widely used, they will be described at this stage and some description given of their effect on processes, even though this can only be done in a qualitative fashion.

A tensile test-piece is shown in Figure 1.1. This is typical of a number of standard test-pieces having a parallel, reduced section for a length that is at least four times the width, w_0. The initial thickness is t_0 and the load on the specimen at any instant, P, is measured by a load cell in the testing machine. In the middle of the specimen, a gauge length l_0 is monitored by an extensometer and at any instant the current gauge length is l and the extension is $\Delta l = l - l_0$. In some tests, a transverse extensometer may also be used to measure the change in width, i.e. $\Delta w = w - w_0$. During the test, load and extension will be recorded in a data acquisition system and a file created; this is then analysed and various material property diagrams can be created. Some of these are described below.

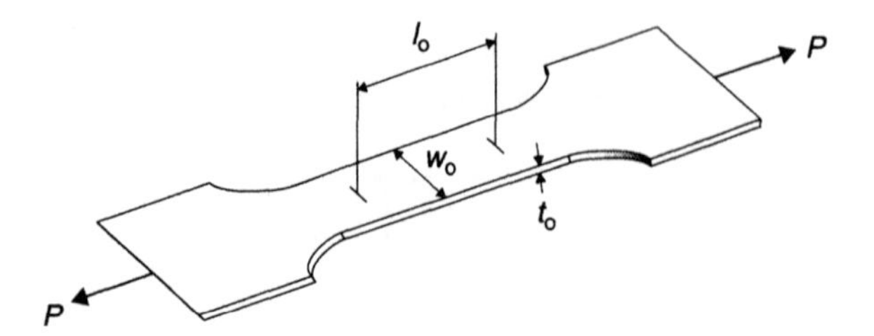

Figure 1.1 Typical tensile test strip.

1.1.1 The load–extension diagram

Figure 1.2 shows a typical load–extension diagram for a test on a sample of drawing quality steel. The elastic extension is so small that it cannot be seen. The diagram does not represent basic material behaviour as it describes the response of the material to a particular process, namely the extension of a tensile strip of given width and thickness. Nevertheless it does give important information. One feature is the *initial yielding load*, P_y, at which plastic deformation commences. Initial yielding is followed by a region in which the deformation in the strip is uniform and the load increases. The increase is due to *strain-hardening*, which is a phenomenon exhibited by most metals and alloys in the soft condition whereby the strength or hardness of the material increases with plastic deformation. During this part of the test, the cross-sectional area of the strip decreases while the length increases; a point is reached when the strain-hardening effect is just balanced by the rate of decrease in area and the load reaches a maximum $P_{max.}$. Beyond this, deformation in the strip ceases to be uniform and a diffuse neck develops in the reduced section; non-uniform extension continues within the neck until the strip fails.

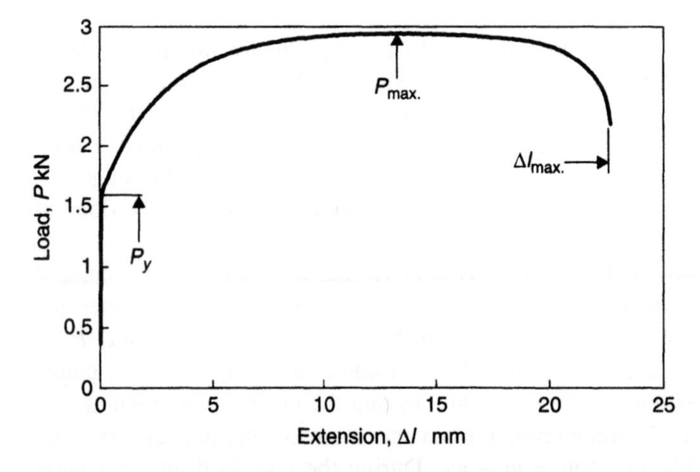

Figure 1.2 Load–extension diagram for a tensile test of a drawing quality sheet steel. The test-piece dimensions are $l_0 = 50$, $w_0 = 12.5$, $t_0 = 0.8$ mm.

The extension at this instant is $\Delta l_{max.}$, and a tensile test property known as the *total elongation* can be calculated; this is defined by

$$E_{Tot.} = \frac{l_{max.} - l_0}{l_0} \times 100\% \tag{1.1}$$

1.1.2 The engineering stress–strain curve

Prior to the development of modern data processing systems, it was customary to scale the load–extension diagram by dividing load by the initial cross-sectional area, $A_0 = w_0 t_0$, and the extension by l_0, to obtain the *engineering stress–strain* curve. This had the advantage that a curve was obtained which was independent of the initial dimensions of the test-piece, but it was still not a true material property curve. During the test, the cross-sectional area will diminish so that the true stress on the material will be greater than the engineering stress. The engineering stress–strain curve is still widely used and a number of properties are derived from it. Figure 1.3(a) shows the engineering stress strain curve calculated from the load, extension diagram in Figure 1.2.

Engineering stress is defined as

$$\sigma_{eng.} = \frac{P}{A_0} \tag{1.2}$$

and *engineering strain* as

$$e_{eng.} = \frac{\Delta l}{l_0} \times 100\% \tag{1.3}$$

In this diagram, the *initial yield stress* is

$$(\sigma_f)_0 = \frac{P_y}{A_0} \tag{1.4}$$

The maximum engineering stress is called the *ultimate tensile strength* or the *tensile strength* and is calculated as

$$TS = \frac{P_{max.}}{A_0} \tag{1.5}$$

As already indicated, this is not the true stress at maximum load as the cross-sectional area is no longer A_0. The elongation at maximum load is called the *maximum uniform elongation*, E_u.

If the strain scale near the origin is greatly increased, the elastic part of the curve would be seen, as shown in Figure 1.3(b). The strain at initial yield, e_y, as mentioned, is very small, typically about 0.1%. The slope of the elastic part of the curve is the *elastic modulus*, also called Youngs modulus:

$$E = \frac{(\sigma_f)_0}{e_y} \tag{1.6}$$

If the strip is extended beyond the elastic limit, permanent plastic deformation takes place; upon unloading, the elastic strain will be recovered and the unloading line is parallel to the initial elastic loading line. There is a residual plastic strain when the load has been removed as shown in Figure 1.3(b).

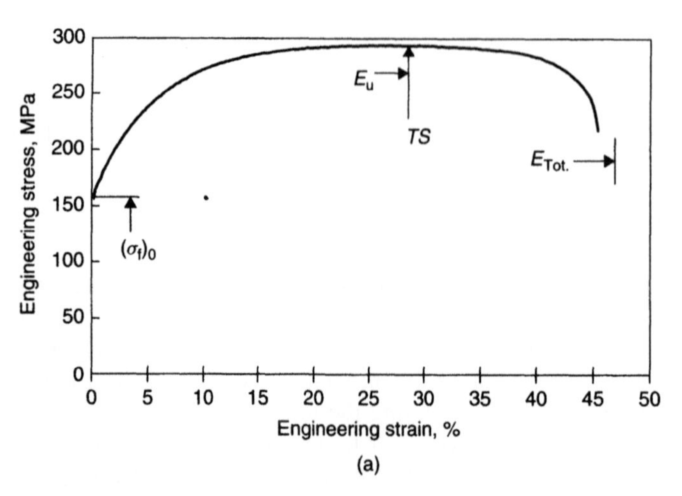

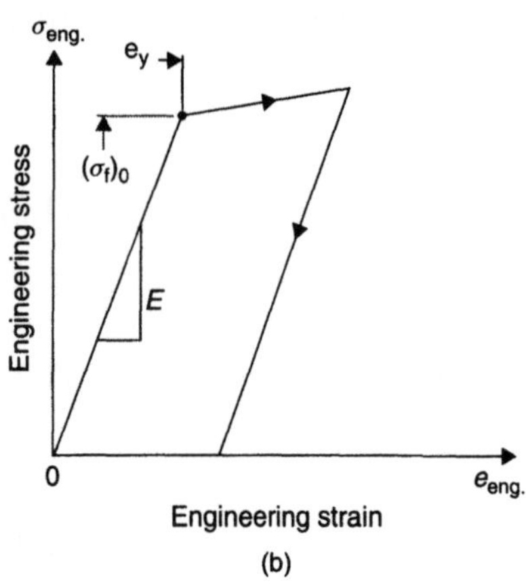

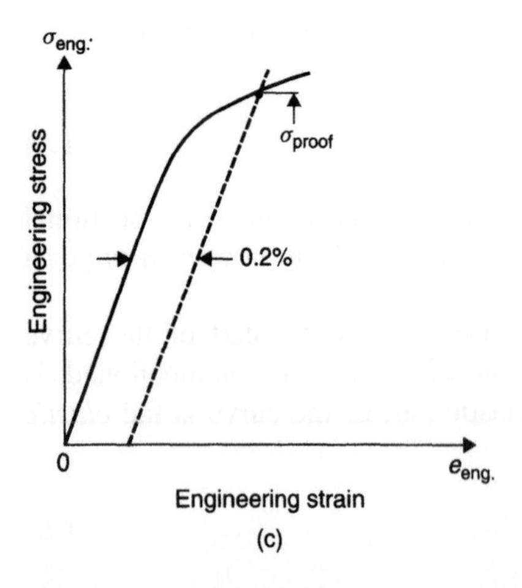

Figure 1.3 (a) Engineering stress–strain curve for the test of drawing quality sheet steel shown in Figure 1.2. (b) Initial part of the above diagram with the strain scale magnified to show the elastic behaviour. (c) Construction used to determine the proof stress in a material with a gradual elastic, plastic transition.

In some materials, the transition from elastic to plastic deformation is not sharp and it is difficult to establish a precise yield stress. If this is the case, a *proof stress* may be quoted. This is the stress to produce a specified small plastic strain – often 0.2%, i.e. about twice the elastic strain at yield. Proof stress is determined by drawing a line parallel to the elastic loading line which is offset by the specified amount, as shown in Figure 1.3(c).

Certain steels are susceptible to *strain ageing* and will display the yield phenomena illustrated in Figure 1.4. This may be seen in some hot-dipped galvanized steels and in bake-hardenable steels used in autobody panels. Ageing has the effect of increasing the initial yielding stress to the *upper yield stress* σ_U; beyond this, yielding occurs in a discontinuous form. In the tensile test-piece, discrete bands of deformation called *Lüder's lines* will traverse the strip under a constant stress that is lower than the upper yield stress; this is known as the *lower yield stress* σ_L. At the end of this discontinuous flow, uniform deformation associated with strain-hardening takes place. The amount of discontinuous strain is called the *yield point elongation* (YPE). Steels that have significant yield point elongation, more than about 1%, are usually unsuitable for forming as they do not deform smoothly and visible markings, called stretcher strains can appear on the part.

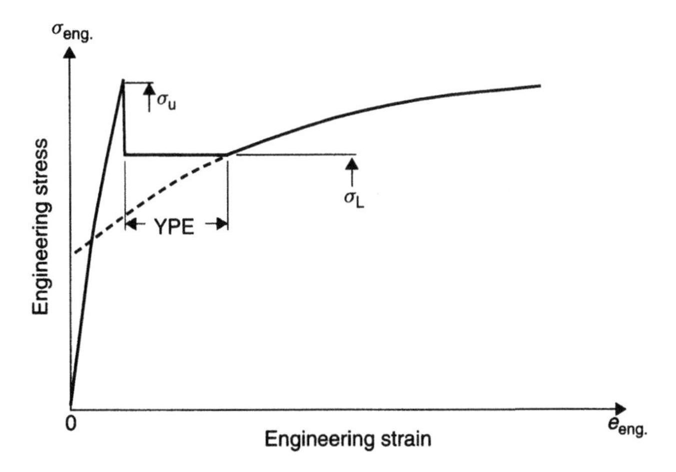

Figure 1.4 Yielding phenomena in a sample of strain aged steel.

1.1.3 The true stress–strain curve

There are several reasons why the engineering stress–strain curve is unsuitable for use in the analysis of forming processes. The 'stress' is based on the initial cross-sectional area of the test-piece, rather than the current value. Also engineering strain is not a satisfactory measure of strain because it is based on the original gauge length. To overcome these disadvantages, the study of forming processes is based on *true stress* and *true strain*; these are defined below.

True stress is defined as

$$\sigma = \frac{P}{A} \tag{1.7}$$

where A is the current cross-sectional area. True stress can be determined from the load–extension diagram during the rising part of the curve, between initial yielding and the maximum load, using the fact that plastic deformation in metals and alloys takes place without any appreciable change in volume. The volume of the gauge section is constant, i.e.

$$A_0 l_0 = Al \tag{1.8}$$

and the true stress is calculated as

$$\sigma = \frac{P}{A_0} \frac{l}{l_0} \tag{1.9}$$

If, during deformation of the test-piece, the gauge length increases by a small amount, dl, a suitable definition of strain is that the *strain increment* is the extension per unit current length, i.e.

$$d\varepsilon = \frac{dl}{l} \tag{1.10}$$

For very small strains, where $l \approx l_0$, the strain increment is very similar to the engineering strain, but for larger strains there is a significant difference. If the straining process continues uniformly in the one direction, as it does in the tensile test, the strain increment can be integrated to give the true strain, i.e.

$$\varepsilon = \int d\varepsilon = \int_{l_0}^{l} \frac{dl}{l} = \ln \frac{l}{l_0} \tag{1.11}$$

The true stress–strain curve calculated from the load–extension diagram above is shown in Figure 1.5. This could also be calculated from the engineering stress–strain diagram using the relationships

$$\sigma = \frac{P}{A} = \frac{P}{A_0}\frac{A_0}{A} = \sigma_{eng.} \frac{l}{l_0} = \sigma_{eng.} \left(1 + \frac{e_{eng.}}{100}\right) \tag{1.12}$$

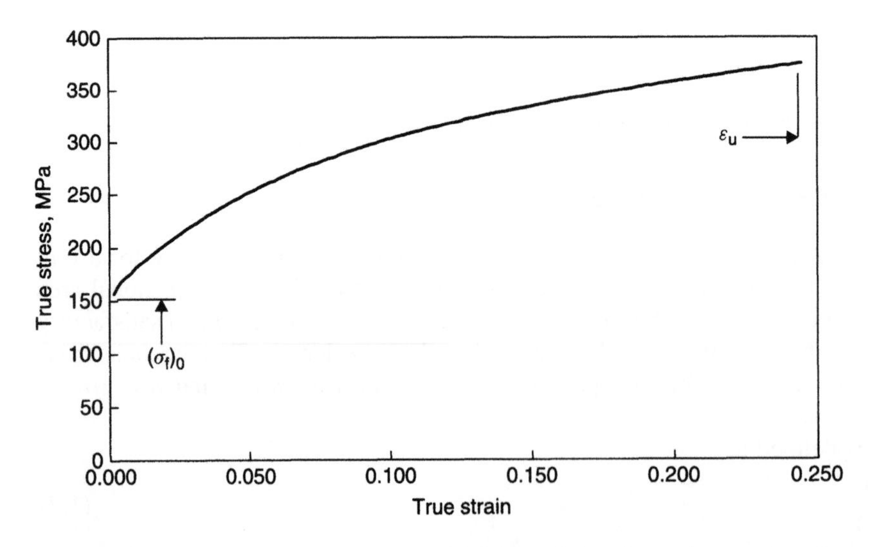

Figure 1.5 The true stress–strain curve calculated from the load–extension diagram for drawing quality sheet steel.

and

$$\varepsilon = \ln\left(1 + \frac{e_{\text{eng.}}}{100}\right) \tag{1.13}$$

It can be seen that the true stress–strain curve does not reach a maximum as strain-hardening is continuous although it occurs at a diminishing rate with deformation. When necking starts, deformation in the gauge length is no longer uniform so that Equation 1.11 is no longer valid. The curve in Figure 1.5 cannot be calculated beyond a strain corresponding to maximum load; this strain is called the *maximum uniform strain*:

$$\varepsilon_{\text{u}} = \ln\left(1 + \frac{E_{\text{u}}}{100}\right) \tag{1.14}$$

If the true stress and strain are plotted on logarithmic scales, as in Figure 1.6, many samples of sheet metal in the soft, annealed condition will show the characteristics of this diagram. At low strains in the elastic range, the curve is approximately linear with a slope of unity; this corresponds to an equation for the elastic regime of

$$\sigma = E\varepsilon \text{ or } \log\sigma = \log E + \log\varepsilon \tag{1.15}$$

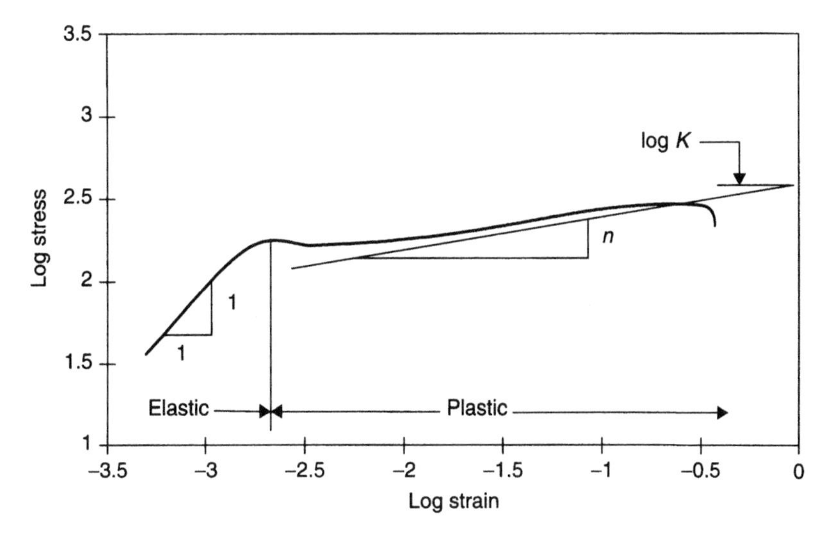

Figure 1.6 True stress–strain from the above diagram plotted in a logarithmic diagram.

At higher strains, the curve shown can be fitted by an equation of the form

$$\sigma = K\varepsilon^n \tag{1.16a}$$

or

$$\log\sigma = \log K + n\log\varepsilon \tag{1.16b}$$

The fitted curve has a slope of n, which is known as the *strain-hardening index*, and an intercept of $\log K$ at a strain of unity, i.e. when $\varepsilon = 1$, or $\log\varepsilon = 0$; K is the *strength coefficient*. The empirical equation or *power law* Equation 1.16(a) is often used to describe the plastic properties of annealed low carbon steel sheet. As may be seen from Figure 1.6, it provides an accurate description, except for the elastic regime and during the first few per cent of plastic strain. Empirical equations of this form are often used to extrapolate the material property description to strains greater than those that can be obtained in the tensile test; this may or may not be valid, depending on the nature of the material.

1.1.4 (Worked example) tensile test properties

The initial gauge length, width and thickness of a tensile test-piece are, 50, 12.5 and 0.80 mm respectively. The initial yield load is 1.791 kN. At a point, A, the load is 2.059 kN and the extension is 1.22 mm. The maximum load is 2.94 kN and this occurs at an extension of 13.55 mm. The test-piece fails at an extension of 22.69 mm.

Determine the following:

initial cross-sectional area, $= 12.5 \times 0.80 = 10\,\text{mm}^2 = 10^{-5}\,\text{m}^2$

initial yield stress, $= \dfrac{1.791 \times 10^3}{10^{-5}} = 179 \times 10^6\ Pa = 179\,\text{MPa}$

tensile strength, $= 2.94 \times 10^3 \div 10^{-5} = 294\,\text{MPa}$

total elongation, $= (22.69/50) \times 100\% = 45.4\%$

true stress at maximum load, $= 294(50 + 13.55)/50 = 374\,\text{MPa}$

maximum uniform strain, $= \ln \dfrac{50 + 13.55}{50} = 0.24$

true stress at A, $= \dfrac{2.059 \times 10^3}{10^{-5}} \times \dfrac{50 + 1.22}{50} = 211\,\text{MPa}$

true strain at A, $= \ln \dfrac{50 + 1.22}{50} = 0.024$

By fitting a power law to two points, point A, and the maximum load point, determine an approximate value of the strain-hardening index and the value of K.

$$n = \frac{\log \sigma_{\text{max.}} - \log \sigma_A}{\log \varepsilon_u - \log \varepsilon_A} = \frac{\log 374 - \log 211}{\log 0.24 - \log 0.024} = 0.25$$

By substitution, $211 = K \times 0.024^{0.25}$, $\therefore K = 536\,\text{MPa}$.
(Note that the maximum uniform strain, 0.24, is close to the value of the strain-hardening index. This can be anticipated, as shown in a later chapter.)

1.1.5 Anisotropy

Material in which the same properties are measured in any direction is termed *isotropic*, but most industrial sheet will show a difference in properties measured in test-pieces aligned, for example, with the rolling, transverse and 45° directions of the coil. This variation is known as *planar anisotropy*. In addition, there can be a difference between the average of properties in the plane of the sheet and those in the through-thickness direction. In tensile tests of a material in which the properties are the same in all directions, one would expect, by symmetry, that the width and thickness strains would be equal; if they are different, this suggests that some *anisotropy* exists.

In materials in which the properties depend on direction, the state of anisotropy is usually indicated by the *R-value*. This is defined as the ratio of width strain, $\varepsilon_w = \ln(w/w_0)$, to thickness strain, $\varepsilon_t = \ln(t/t_0)$. In some cases, the thickness strain is measured directly, but it may be calculated also from the length and width measurements using the constant volume assumption, i.e.

$$wtl = w_0 t_0 l_0$$

or

$$\frac{t}{t_0} = \frac{w_0 l_0}{wl}$$

The R-value is therefore,

$$R = \frac{\ln \dfrac{w}{w_0}}{\ln \dfrac{w_0 l_0}{wl}} \tag{1.17}$$

If the change in width is measured during the test, the R-value can be determined continuously and some variation with strain may be observed. Often measurements are taken at a particular value of strain, e.g. at $e_{\text{eng.}} = 15\%$. The direction in which the R-value is measured is indicated by a suffix, i.e. R_0, R_{45} and R_{90} for tests in the rolling, diagonal and transverse directions respectively. If, for a given material, these values are different, the sheet is said to display *planar anisotropy* and the most common description of this is

$$\Delta R = \frac{R_0 + R_{90} - 2R_{45}}{2} \tag{1.18}$$

which may be positive or negative, although in steels it is usually positive.

If the measured R-value differs from unity, this shows a difference between average in-plane and through-thickness properties which is usually characterized by the *normal plastic anisotropy ratio*, defined as

$$\overline{R} = \frac{R_0 + 2R_{45} + R_{90}}{4} \tag{1.19}$$

The term 'normal' is used here in the sense of properties 'perpendicular' to the plane of the sheet.

1.1.6 Rate sensitivity

For many materials at room temperature, the properties measured will not vary greatly with small changes in the speed at which the test is performed. The property most sensitive to rate of deformation is the lower yield stress and therefore it is customary to specify the cross-head speed of the testing machine – typically about 25 mm/minute.

If the cross-head speed, v, is suddenly changed by a factor of 10 or more during the uniform deformation region of a tensile test, a small jump in the load may be observed as shown in Figure 1.7. This indicates some *strain-rate sensitivity* in the material that can be described by the exponent, m, in the equation

$$\sigma = K\varepsilon^n \dot{\varepsilon}^m \tag{1.20}$$

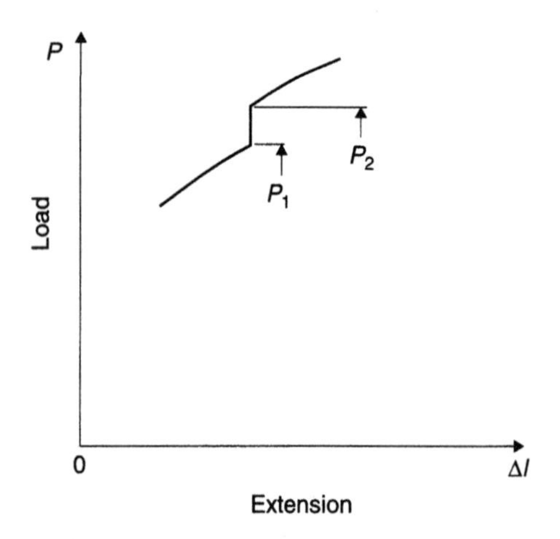

Figure 1.7 Part of a load–extension diagram showing the jump in load following a sudden increase in extension rate.

The *strain rate* is

$$\dot{\varepsilon} = \frac{v}{L} \tag{1.21}$$

where L denotes the length of the parallel reduced section of the test-piece. The exponent m is calculated from load and cross-head speed before and after the speed change, denoted by suffixes 1 and 2 respectively; i.e.

$$m = \frac{\log(P_1/P_2)}{\log(v_1/v_2)} \tag{1.21}$$

1.2 Effect of properties on forming

It is found that the way in which a given sheet behaves in a forming process will depend on one or more general characteristics. Which of these is important will depend on the particular process and by studying the mechanics equations governing the process it is often possible to predict those properties that will be important. This assumes that the property has a fundamental significance, but as mentioned above, not all the properties obtained from the tensile test will fall into this category.

The general attributes of material behaviour that affect sheet metal forming are as follows.

1.2.1 Shape of the true stress–strain curve

The important aspect is strain-hardening. The greater the strain-hardening of the sheet, the better it will perform in processes where there is considerable stretching; the straining will be more uniformly distributed and the sheet will resist tearing when strain-hardening is high. There are a number of indicators of strain-hardening and the strain-hardening index, n, is the most precise. Other measures are the *tensile/yield ratio*, $TS/(\sigma_f)_0$, the total elongation, $E_{Tot.}$ and the maximum uniform strain, ε_u; the higher these are, the greater is the strain-hardening.

The importance of the initial yield strength, as already mentioned, is related to the strength of the formed part and particularly where lightweight construction is desired, the higher the yield strength, the more efficient is the material. Yield strength does not directly affect forming behaviour, although usually higher strength sheet is more difficult to form; this is because other properties change in an adverse manner as the strength increases.

The elastic modulus also affects the performance of the formed part and a higher modulus will give a stiffer component, which is usually an advantage. In terms of forming, the modulus will affect the springback. A lower modulus gives a larger springback and usually more difficulty in controlling the final dimensions. In many cases, the springback will increase with the ratio of yield stress to modulus, $(\sigma_f)_0/E$, and higher strength sheet will also have greater springback.

1.2.2 Anisotropy

If the magnitude of the planar anisotropy parameter, ΔR, is large, either, positive or negative, the orientation of the sheet with respect to the die or the part to be formed will be important; in circular parts, asymmetric forming and earing will be observed. If the normal anisotropy ratio $\overline{R}$ is greater than unity it indicates that in the tensile test the width strain is greater than the thickness strain; this may be associated with a greater strength in the through-thickness direction and, generally, a resistance to thinning. Normal anisotropy $\overline{R}$ also has more subtle effects. In drawing deep parts, a high value allows deeper parts to be drawn. In shallow, smoothly-contoured parts such as autobody outer panels, a higher value of $\overline{R}$ may reduce the chance of wrinkling or ripples in the part. Other factors such as inclusions, surface topography, or fracture properties may also vary with orientation; these would not be indicated by the R-value which is determined from plastic properties.

1.2.3 Fracture

Even in ductile materials, tensile processes can be limited by sudden fracture. The fracture characteristic is not given by total elongation but is indicated by the cross-sectional area of the fracture surface after the test-piece has necked and failed. This is difficult to measure in thin sheet and consequently problems due to fracture may not be properly recognized.

1.2.4 Homogeneity

Industrial sheet metal is never entirely homogeneous, nor free from local defects. Defects may be due to variations in composition, texture or thickness, or exist as point defects such as inclusions. These are difficult to characterize precisely. Inhomogeneity is not indicated by a single tensile test and even with repeated tests, the actual volume of material being tested is small, and non-uniformities may not be adequately identified.

1.2.5 Surface effects

The roughness of sheet and its interaction with lubricants and tooling surfaces will affect performance in a forming operation, but will not be measured in the tensile test. Special tests exist to explore surface properties.

1.2.6 Damage

During tensile plastic deformation, many materials suffer damage at the microstructural level. The rate at which this damage progresses varies greatly with different materials. It may be indicated by a diminution in strain-hardening in the tensile test, but as the rate of damage accumulation depends on the stress state in the process, tensile data may not be indicative of damage in other stress states.

1.2.7 Rate sensitivity

As mentioned, the rate sensitivity of most sheet is small at room temperature; for steel it is slightly positive and for aluminium, zero or slightly negative. Positive rate sensitivity usually improves forming and has an effect similar to strain-hardening. As well as being indicated by the exponent m, it is also shown by the amount of extension in the tensile test-piece after maximum load and necking and before failure, i.e. $E_{\text{Total}} - E_{\text{u}}$, increases with increasing rate sensitivity.

1.2.8 Comment

It will be seen that the properties that affect material performance are not limited to those that can be measured in the tensile test or characterized by a single value. Measurement of homogeneity and defects may require information on population, orientation and spatial distribution.

Many industrial forming operations run very close to a critical limit so that small changes in material behaviour give large changes in failure rates. When one sample of material will run in a press and another will not, it is frequently observed that the materials cannot be distinguished in terms of tensile test properties. This may mean that one or two tensile tests are insufficient to characterize the sheet or that the properties governing the performance are only indicated by some other test.

1.3 Other mechanical tests

As mentioned, the tensile test is the most widely used mechanical test, but there are many other mechanical tests in use. For example, in the study of bulk forming processes such as forging and extrusion, compression tests are common, but these are not suitable for sheet. Some tests appropriate for sheet are briefly mentioned below:

- *Springback.* The elastic properties of sheet are not easily measured in routine tensile tests, but they do affect springback in parts. For this reason a variety of springback tests have been devised where the sheet is bent over a former and then released.
- *Hardness tests.* An indenter is pressed into the sheet under a controlled load and the size of the impression measured. This will give an approximate measure of the hardness of the sheet – the smaller the impression, the greater the hardness. Empirical relations allow hardness readings to be converted to 'yield strength'. For strain-hardening materials, this yield strength will be roughly the average of initial yield and ultimate tensile strength. The correlation is only approximate, but hardness tests can usefully distinguish one grade of sheet from another.

- *Hydrostatic bulging test.* In this test a circular disc is clamped around the edge and bulged to a domed shape by fluid pressure. From measurement of pressure, curvature and membrane or thickness strain at the pole, a true stress–strain curve under equal biaxial tension can be obtained. The advantage of this test is that for materials that have little strain-hardening, it is possible to obtain stress–strain data over a much larger strain range than is possible in the tensile test.
- *Simulative tests.* A number of tests have been devised in which sheet is deformed in a particular process using standard tooling. Examples include drawing a cup, stretching over a punch and expanding a punched hole. The principles of these tests are covered in later chapters.

1.4 Exercises

Ex. 1.1 A tensile specimen is cut from a sheet of steel of 1 mm thickness. The initial width is 12.5 mm and the gauge length is 50 mm.

(a) The initial yield load is 2.89 kN and the extension at this point is 0.0563 mm. Determine the initial yield stress and the elastic modulus.
(b) When the extension is 15%, the width of the test-piece is 11.41. Determine the R-value.

[Ans: (a): 231 MPa, 205 GPa; (b) 1.88]

Ex. 1.2 At 4% and 8% elongation, the loads on a tensile test-piece of half-hard aluminium alloy are 1.59 kN and 1.66 kN respectively. The test-piece has an initial width of 10 mm, thickness of 1.4 mm and gauge length of 50 mm. Determine the K and n values.
[Ans: 174 MPa, 0.12]

Ex. 1.3 The K, n and m values for a stainless steel sheet are 1140 MPa, 0.35 and 0.01 respectively. A test-piece has initial width, thickness and gauge length of 12.5, 0.45 and 50 mm respectively. Determine the increase in load when the extension is 10% and the extension rate of the gauge length is increased from 0.5 to 50 mm/minute.
[Ans: 0.27 kN]

Ex. 1.4 The following data pairs (load kN; extension mm) were obtained from the plastic part of a load-extension file for a tensile test on an extra deep drawing quality steel sheet of 0.8 mm thickness. The initial test-piece width was 12.5 mm and the gauge length 50 mm.

1.57, 0.080; 1.90, 0.760; 2.24, 1.85; 2.57, 3.66; 2.78, 5.84; 2.90, 8.92
2.93,11.06; 2.94,13.49; 2.92, 16.59; 2.86, 19.48; 2.61, 21.82; 2.18, 22.69

Obtain engineering stress–strain, true stress, strain and log stress, log strain curves. From these determine; initial yield stress, ultimate tensile strength, true strain at maximum load, total elongation and the strength coefficient, K, and strain-hardening index, n.
[Ans: 156 MPa, 294 MPa, 0.24, 45%, 530 MPa, 0.24]

2 Sheet deformation processes

2.1 Introduction

In Chapter 1, the appropriate definitions for stress and strain in tensile deformation were introduced. The purpose now is to indicate how the true stress–strain curve derived from a tensile test can be applied to other deformation processes that may occur in typical sheet forming operations.

A common feature of many sheet forming processes is that the stress perpendicular to the surface of the sheet is small, compared with the stresses in the plane of the sheet (the membrane stresses). If we assume that this *normal stress* is zero, a major simplification is possible. Such a process is called *plane stress* deformation and the theory of yielding for this process is described in this chapter. There are cases in which the through-thickness or normal stress cannot be neglected and the theory of yielding in a three-dimensional stress state is described in an appendix.

The tensile test is of course a plane stress process, *uniaxial tension*, and this is now reviewed as an example of plane stress deformation.

2.2 Uniaxial tension

We consider an element in a tensile test-piece in uniaxial deformation and follow the process from an initial small change in shape. Up to the maximum load, the deformation is uniform and the element chosen can be large and, in Figure 2.1, we consider the whole gauge section. During deformation, the faces of the element will remain perpendicular to each other as it is, by inspection, a *principal element*, i.e. there is no shear strain associated with the *principal directions*, 1, 2 and 3, along the axis, across the width and through the thickness, respectively.

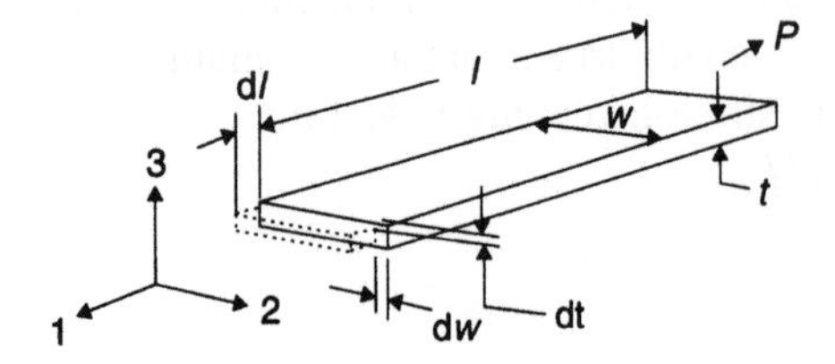

Figure 2.1 The gauge element in a tensile test-piece showing the principal directions.

2.2.1 Principal strain increments

During any small part of the process, the *principal strain increment* along the tensile axis is given by Equation 1.10 and is

$$d\varepsilon_1 = \frac{dl}{l} \tag{2.1}$$

i.e. the increase in length per unit current length.

Similarly, across the strip and in the through-thickness direction the strain increments are

$$d\varepsilon_2 = \frac{dw}{w} \quad \text{and} \quad d\varepsilon_3 = \frac{dt}{t} \tag{2.2}$$

2.2.2 Constant volume (incompressibility) condition

It has been mentioned that plastic deformation occurs at constant volume so that these strain increments are related in the following manner. With no change in volume, the differential of the volume of the gauge region will be zero, i.e.

$$d(lwt) = d(l_o w_0 t_o) = 0$$

and we obtain

$$dl \times wt + dw \times lt + dt \times lw = 0$$

or, dividing by *lwt*,

$$\frac{dl}{l} + \frac{dw}{w} + \frac{dt}{t} = 0$$

i.e.

$$d\varepsilon_1 + d\varepsilon_2 + d\varepsilon_3 = 0 \tag{2.3}$$

Thus for constant volume deformation, the sum of the principal strain increments is zero.

2.2.3 Stress and strain ratios (isotropic material)

If we now restrict the analysis to isotropic materials, where identical properties will be measured in all directions, we may assume from symmetry that the strains in the width and thickness directions will be equal in magnitude and hence, from Equation 2.3,

$$d\varepsilon_2 = d\varepsilon_3 = -\frac{1}{2} d\varepsilon_1$$

(In the previous chapter we considered the case in which the material was anisotropic where $d\varepsilon_2 = R d\varepsilon_3$ and the R-value was not unity. We can develop a general theory for anisotropic deformation, but this is not necessary at this stage.)

We may summarize the tensile test process for an isotropic material in terms of the strain increments and stresses in the following manner:

$$d\varepsilon_1 = \frac{dl}{l}; \qquad d\varepsilon_2 = -\frac{1}{2}d\varepsilon_1; \qquad d\varepsilon_3 = -\frac{1}{2}d\varepsilon_1 \tag{2.4a}$$

and

$$\sigma_1 = \frac{P}{A}; \qquad \sigma_2 = 0; \qquad \sigma_3 = 0 \tag{2.4b}$$

2.2.4 True, natural or logarithmic strains

It may be noted that in the tensile test the following conditions apply:

- the principal strain increments all increase smoothly in a constant direction, i.e. $d\varepsilon_1$ always increases positively and does not reverse; this is termed a *monotonic* process;
- during the uniform deformation phase of the tensile test, from the onset of yield to the maximum load and the start of diffuse necking, the ratio of the principal strains remains constant, i.e. the process is *proportional;* and
- the principal directions are fixed in the material, i.e. the direction 1 is always along the axis of the test-piece and a material element does not rotate with respect to the principal directions.

If, and only if, these conditions apply, we may safely use the integrated or large strains defined in Chapter 1. For uniaxial deformation of an isotropic material, these strains are

$$\varepsilon_1 = \ln\frac{l}{l_0}; \qquad \varepsilon_2 = \ln\frac{w}{w_0} = -\frac{1}{2}\varepsilon_1; \qquad \varepsilon_3 = \ln\frac{t}{t_o} = -\frac{1}{2}\varepsilon_1 \tag{2.5}$$

2.3 General sheet processes (plane stress)

In contrast with the tensile test in which two of the principal stresses are zero, in a typical sheet process most elements will deform under membrane stresses σ_1 and σ_2, which are both non-zero. The third stress, σ_3, perpendicular to the surface of the sheet is usually quite small as the contact pressure between the sheet and the tooling is generally very much lower than the yield stress of the material. As indicated above, we will make the simplifying assumption that it is zero and assume *plane stress deformation*, unless otherwise stated. If we also assume that the same conditions of proportional, monotonic deformation apply as for the tensile test, then we can develop a simple theory of plastic deformation of sheet that is reasonably accurate. We can illustrate these processes for an element as shown in Figure 2.2(a) for the uniaxial tension and Figure 2.2(b) for a general plane stress sheet process.

2.3.1 Stress and strain ratios

It is convenient to describe the deformation of an element, as in Figure 2.2(b), in terms of either the strain ratio β or the stress ratio α. For a proportional process, which is the only kind we are considering, both will be constant. The usual convention is to define the

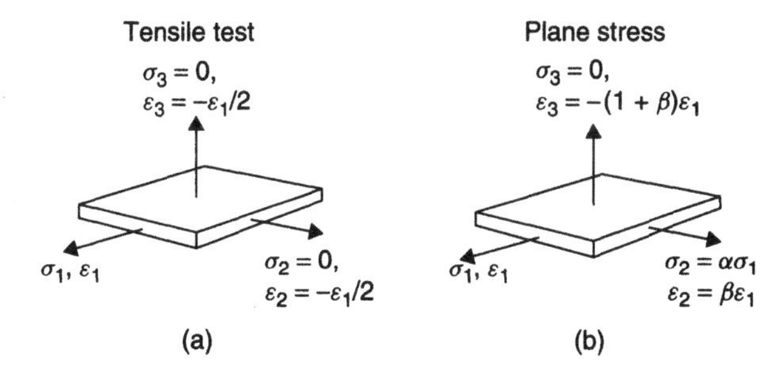

Figure 2.2 Principal stresses and strains for elements deforming in (a) uniaxial tension and (b) a general plane stress sheet process.

principal directions so that $\sigma_1 > \sigma_2$ and the third direction is perpendicular to the surface where $\sigma_3 = 0$. The deformation mode is thus:

$$\varepsilon_1; \qquad \varepsilon_2 = \beta\varepsilon_1; \qquad \varepsilon_3 = -(1+\beta)\varepsilon_1 \tag{2.6}$$

$$\sigma_1; \qquad \sigma_2 = \alpha\sigma_1; \qquad \sigma_3 = 0$$

The constant volume condition is used to obtain the third principal strain. Integrating the strain increments in Equation 2.3 shows that this condition can be expressed in terms of the true or natural strains:

$$\varepsilon_1 + \varepsilon_2 + \varepsilon_3 = 0 \tag{2.7}$$

i.e. the sum of the natural strains is zero.

For uniaxial tension, the strain and stress ratios are $\beta = -1/2$ and $\alpha = 0$.

2.4 Yielding in plane stress

The stresses required to yield a material element under plane stress will depend on the current hardness or strength of the sheet and the stress ratio α. The usual way to define the strength of the sheet is in terms of the current flow stress σ_f. The flow stress is the stress at which the material would yield in simple tension, i.e. if $\alpha = 0$. This is illustrated in the true stress–strain curve in Figure 2.3. Clearly σ_f depends on the amount of deformation to which the element has been subjected and will change during the process. For the moment, we shall consider only one instant during deformation and, knowing the current value of σ_f the objective is to determine, for a given value of α, the values of σ_1 and σ_2 at which the element will yield, or at which plastic flow will continue for a small increment. We consider here only the instantaneous conditions in which the strain increment is so small that the flow stress can be considered constant. In Chapter 3 we extend this theory for continuous deformation.

There are a number of theories available for predicting the stresses under which a material element will deform plastically. Each theory is based on a different hypothesis about material behaviour, but in this work we shall only consider two common models and apply them to the plane stress process described by Equations 2.6. Over the years, many researchers have conducted experiments to determine how materials yield. While no single theory agrees exactly with experiment, for isotropic materials either of the models presented here are sufficiently accurate for approximate models.

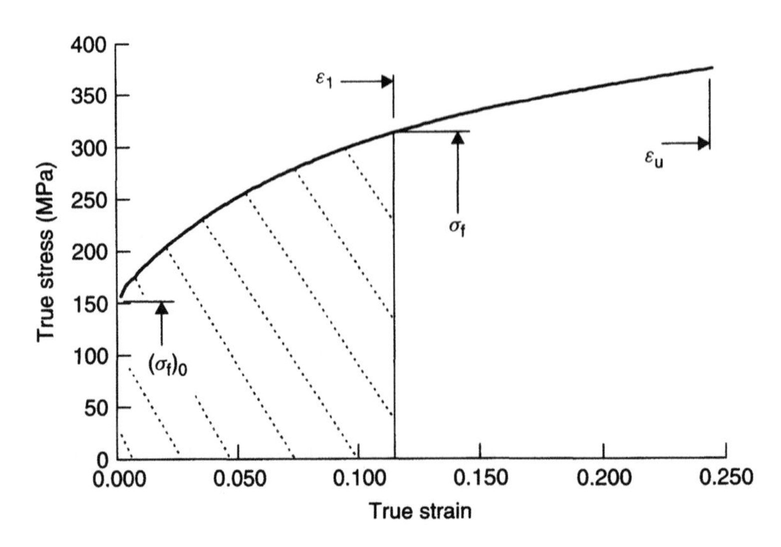

Figure 2.3 Diagram showing the current flow stress in an element after some strain in a tensile test.

With hindsight, common yielding theories can be anticipated from knowledge of the nature of plastic deformations in metals. These materials are polycrystalline and plastic flow occurs by slip on crystal lattice planes when the shear stress reaches a critical level. To a first approximation, this slip which is associated with dislocations in the lattice is insensitive to the normal stress on the slip planes. It may be anticipated then that yielding will be associated with the shear stresses on the element and is not likely to be influenced by the average stress or pressure. It is appropriate to define these terms more precisely.

2.4.1 Maximum shear stress

On the faces of the principal element on the left-hand side of Figure 2.4, there are no shear stresses. On a face inclined at any other angle, both normal and shear stresses will act. On faces of different orientation it is found that the shear stresses will locally reach a maximum for three particular directions; these are the *maximum shear stress planes* and are illustrated in Figure 2.4. They are inclined at 45° to the principal directions and the *maximum shear stresses* can be found from the Mohr circle of stress, Figure 2.5. Normal stresses also act on these maximum shear stress planes, but these have not been shown in the diagram.

The three maximum shear stresses for the element are

$$\tau_1 = \frac{\sigma_1 - \sigma_2}{2}; \qquad \tau_2 = \frac{\sigma_2 - \sigma_3}{2}; \qquad \tau_3 = \frac{\sigma_3 - \sigma_1}{2} \tag{2.8}$$

From the discussion above, it might be anticipated that yielding would be dependent on the shear stresses in an element and the current value of the flow stress; i.e. that a yielding condition might be expressed as

$$f(\tau_1, \tau_2, \tau_3) = \sigma_f$$

We explore this idea below.

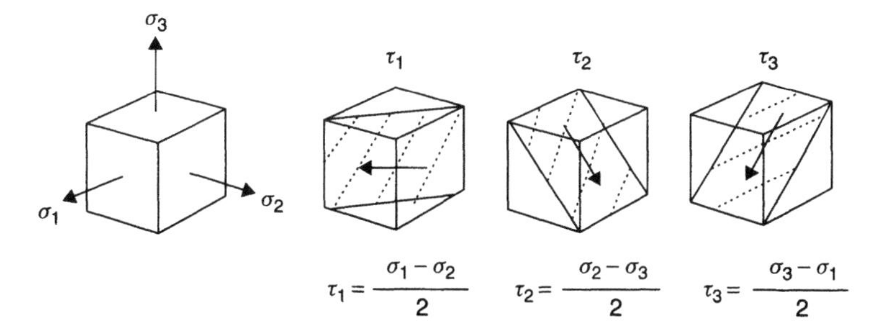

Figure 2.4 A principal element and the three maximum shear planes and stresses.

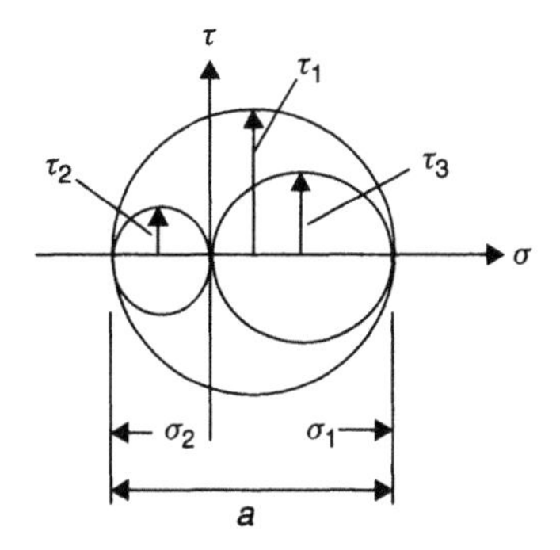

Figure 2.5 The Mohr circle of stress showing the maximum shear stresses.

2.4.2 The hydrostatic stress

The hydrostatic stress is the average of the principal stresses and is defined as

$$\sigma_h = \frac{\sigma_1 + \sigma_2 + \sigma_3}{3} \tag{2.9}$$

It can be considered as three equal components acting in all directions on the element as shown in Figure 2.6.

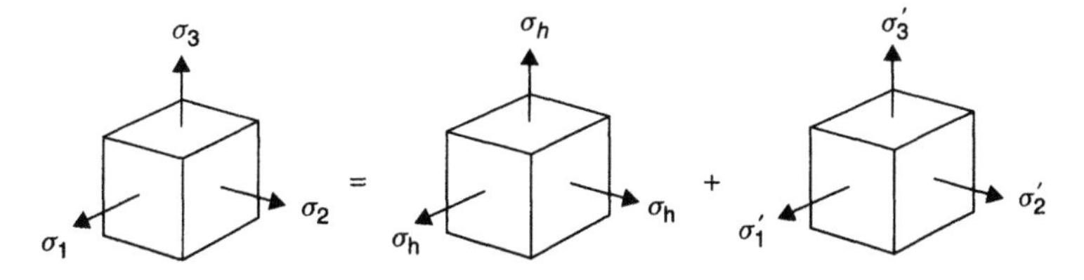

Figure 2.6 A principal element showing how the principal stress state can be composed of hydrostatic and deviatoric components.

Hydrostatic stress is similar to the hydrostatic pressure p in a fluid, except that, by convention in fluid mechanics, p is positive for compression, while in the mechanics of solids, a compressive stress is negative, hence,

$$\sigma_h = -p$$

As indicated above, it may be anticipated that this part of the stress system will not contribute to deformation in a material that deforms at constant volume.

2.4.3 The deviatoric or reduced component of stress

In Figure 2.6, the components of stress remaining after subtracting the hydrostatic stress have a special significance. They are called the *deviatoric*, or *reduced* stresses and are defined by

$$\sigma_1' = \sigma_1 - \sigma_h; \qquad \sigma_2' = \sigma_2 - \sigma_h; \qquad \sigma_3' = \sigma_3 - \sigma_h \tag{2.10a}$$

In plane stress, this may also be written in terms of the stress ratio, i.e.

$$\sigma_1' = \frac{2-\alpha}{3}\sigma_1; \qquad \sigma_2' = \frac{2\alpha-1}{3}\sigma_1; \qquad \sigma_3' = -\left(\frac{1+\alpha}{3}\right)\sigma_1 \tag{2.10b}$$

The reduced or deviatoric stress is the difference between the principal stress and the hydrostatic stress.

The theory of yielding and plastic deformation can be described simply in terms of either of these components of the state of stress at a point, namely, the maximum shear stresses, or the deviatoric stresses.

2.4.4 The Tresca yield condition

One possible hypothesis is that yielding would occur when the greatest maximum shear stress reaches a critical value. In the tensile test where $\sigma_2 = \sigma_3 = 0$, the greatest maximum shear stress at yielding is $\tau_{crit.} = \sigma_f/2$. Thus in this theory, the *Tresca yield criterion,* yielding would occur in any process when

$$\frac{\sigma_{max.} - \sigma_{min.}}{2} = \frac{\sigma_f}{2}$$

or, as usually stated,

$$|\sigma_{max.} - \sigma_{min.}| = \sigma_f \tag{2.11}$$

In plane stress, using the notation here, σ_1 will be the maximum stress and, $\sigma_3 = 0$, the through-thickness stress. The minimum stress will be either σ_3 if σ_2 is positive, or, if σ_2 is negative, it will be σ_2. In all cases, the diameter a of the Mohr circle of stress in Figure 2.5 will be equal to σ_f.

The Tresca yield criterion in plane stress can be illustrated graphically by the hexagon shown in Figure 2.7. The hexagon is the locus of a point P that indicates the stress state at yield as the stress ratio α changes. In a work-hardening material, this locus will expand as σ_f increases, but here we consider only the instantaneous conditions where the flow stress is constant.

2.4.5 The von Mises yield condition

The other widely used criterion is that yielding will occur when the root-mean-square value of the maximum shear stresses reaches a critical value. Several names have been associated with this criterion and here we shall call it the *von Mises* yield theory.

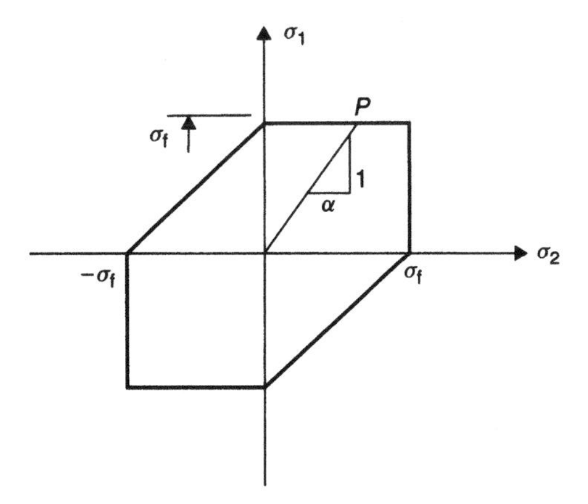

Figure 2.7 Yield locus for plane stress for the Tresca yield criterion.

Bearing in mind that in the tensile test at yield, two of the maximum shear stresses will have the value of $\sigma_f/2$, while the third is zero, this criterion can be expressed mathematically as

$$\sqrt{\frac{\tau_1^2 + \tau_2^2 + \tau_3^2}{3}} = \sqrt{\frac{2\,(\sigma_f/2)^2}{3}}$$

or

$$\sqrt{2(\tau_1^2 + \tau_2^2 + \tau_3^2)} = \sigma_f \tag{2.12a}$$

Substituting the principal stresses for the maximum shear stresses from Equation 2.8, the yielding condition can be expressed also as

$$\sqrt{\frac{1}{2}\{(\sigma_1 - \sigma_2)^2 + (\sigma_2 - \sigma_3)^2 + (\sigma_3 - \sigma_1)^2\}} = \sigma_f \tag{2.12b}$$

By substituting for the deviatoric stresses, i.e.

$$\sigma_1' = (2\sigma_1 - \sigma_2 - \sigma_3)/3 \text{ etc.}$$

the yield condition can be written as

$$\sqrt{\frac{3}{2}(\sigma_1'^2 + \sigma_2'^2 + \sigma_3'^2)} = \sigma_f \tag{2.12c}$$

For the plane stress state specified in Equation 2.6, the criterion is

$$\sqrt{\sigma_1^2 - \sigma_1\sigma_2 + \sigma_2^2} = \left(\sqrt{1 - \alpha + \alpha^2}\right)\sigma_1 = \sigma_f \tag{2.12d}$$

In the principal stress space, this is an ellipse as shown in Figure 2.8.

It is reiterated that both the above theories apply only to isotropic material and they are a reasonable approximation to experimental observations. Although there are major differences in the mathematical form of these two criteria, the values of stress predicted for any given value of α will not differ by more than 15%. In the Mohr circle of stress, the diameter of the largest circle, a, in Figure 2.5 will be in the range

$$\sigma_f \le a \le \frac{2}{\sqrt{3}}\sigma_f = (1.15\sigma_f)$$

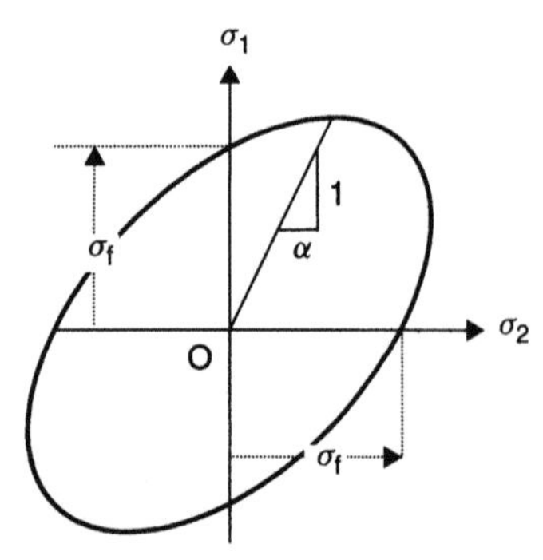

Figure 2.8 Yield locus for plane stress for a von Mises yield condition.

2.5 The flow rule

A yield theory allows one to predict the values of stress at which a material element will deform plastically in plane stress, provided the ratio of the stresses in the plane of the sheet and the flow stress of the material are known. In the study of metal forming processes, we will also need to be able to determine what strains will be associated with the stress state when the element deforms. In elastic deformation, there is a one-to-one relation between stress and strain; i.e. if we know the stress state we can determine the strain state and vice versa. We are already aware of this, because in the experimental study of elastic structures, stresses are determined by strain gauges. This is not possible in the plastic regime. A material element may be at a yielding stress state, i.e. the stresses satisfy the yield condition, but there may be no change in shape. Alternatively, an element in which the stress state is a yielding one, may undergo some small increment of strain that is determined by the displacements of the boundaries; i.e. the magnitude of the deformation increment is determined by the movement of the boundaries and not by the stresses. However, what can be predicted if deformation occurs is the *ratio* of the strain increments; this does depend on the stress state.

Again, with hindsight, the relationship between the stress and strain ratios can be anticipated by considering the nature of flow. In the tensile test, the stresses are in the ratio

$$1 : 0 : 0$$

and the strains in the ratio

$$1 : -1/2 : -1/2$$

so it is not simply a matter of the stresses and strains being in the same ratio. The appropriate relation for general deformation is not, therefore, an obvious one, but can be found by resolving the stress state into the two components, namely the *hydrostatic stress* and the *reduced* or *deviatoric stresses* that have been defined above.

2.5.1 The Levy–Mises flow rule

As shown in Figure 2.6, the deviatoric or reduced stress components, together with the hydrostatic components, make up the actual stress state. As the hydrostatic stress is unlikely to influence deformation in a solid that deforms at constant volume, it may be surmised that it is the deviatoric components that will be the ones associated with the shape change. This is the hypothesis of the *Levy–Mises Flow Rule*. This states that the ratio of the strain increments will be the same as the ratio of the deviatoric stresses, i.e.

$$\frac{d\varepsilon_1}{\sigma_1'} = \frac{d\varepsilon_2}{\sigma_2'} = \frac{d\varepsilon_3}{\sigma_3'} \tag{2.13a}$$

or

$$\frac{d\varepsilon_1}{2-\alpha} = \frac{d\varepsilon_2}{2\alpha - 1} = \frac{d\varepsilon_3}{-(1+\alpha)} \tag{2.13b}$$

If a material element is deforming in a plane stress, proportional process, as described by Equation 2.6, then Equation 2.13(b) can be integrated and expressed in terms of the natural or true strains, i.e.

$$\frac{\varepsilon_1}{2-\alpha} = \frac{\varepsilon_2}{2\alpha - 1} = \frac{\beta\varepsilon_1}{2\alpha - 1} = \frac{\varepsilon_3}{-(1+\alpha)} = \frac{-(1+\beta)\varepsilon_1}{-(1+\alpha)} \tag{2.13c}$$

2.5.2 Relation between the stress and strain ratios

From the above, we obtain the relation between the stress and strain ratios:

$$\alpha = \frac{2\beta + 1}{2 + \beta} \quad \text{and} \quad \beta = \frac{2\alpha - 1}{2 - \alpha} \tag{2.14}$$

It may be seen that while the flow rule gives the relation between the stress and strain ratios, it does not indicate the magnitude of the strains. If the element deforms under a given stress state (i.e. α is known) the ratio of the strains can be found from Equation 2.13, or 2.14. The relationship can be illustrated for different load paths as shown in Figure 2.9; the small arrows show the ratio of the principal strain increments and the lines radiating from the origin indicate the loading path on an element. It may be seen that each of these strain increment vectors is perpendicular to the von Mises yield locus. (It is possible to predict this from considerations of energy or work.)

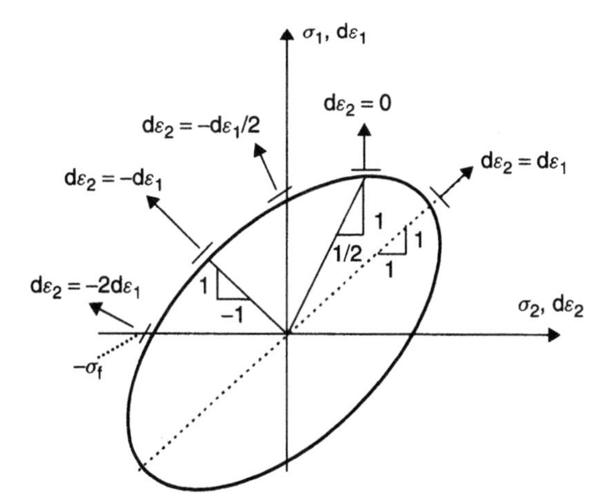

Figure 2.9 Diagram showing the strain increment components for different stress states around the von Mises yield locus.

2.5.3 (Worked example) stress state

The current flow stress of a material element is 300 MPa. In a deformation process, the principal strain increments are 0.012 and 0.007 in the 1 and 2 directions respectively. Determine the principal stresses associated with this in a plane stress process.

Solution

$d\varepsilon_2 = \beta d\varepsilon_1$, $\therefore \beta = d\varepsilon_2/d\varepsilon_1 = 0.007/0.012 = 0.583$.

$\therefore \alpha = (2\beta + 1)/(2 + \beta) = 0.839$

$\therefore \sigma_1 = \sigma_f/\sqrt{(1 - \alpha + \alpha^2)} = 323\,\text{MPa}$; $\quad \sigma_2 = \alpha.\sigma_1 = 271\,\text{MPa}$

2.6 Work of plastic deformation

If we consider a unit principal element as shown in Figure 2.10, then for a small deformation, each side of the unit cube will move by an amount,

$$1 \times d\varepsilon_1; \quad 1 \times d\varepsilon_2; \text{ etc.}$$

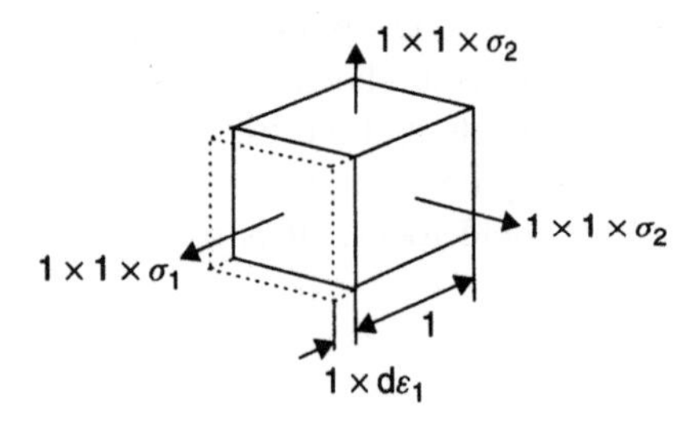

Figure 2.10 Diagram of a principal element of unit side, showing the force acting on a face and its displacement during a small deformation.

and as the force on each face is $\sigma_1 \times 1 \times 1$, etc., the work done in deforming the unit element is

$$\frac{dW}{\text{vol.}} = \sigma_1\, d\varepsilon_1 + \sigma_2\, d\varepsilon_2 + \sigma_3\, d\varepsilon_3 \tag{2.15a}$$

For a plane stress process, this becomes

$$\frac{W}{\text{vol.}} = \int_0^{\varepsilon_1} \sigma_1\, d\varepsilon_1 + \int_0^{\varepsilon_2} \sigma_2\, d\varepsilon_2 \tag{2.15b}$$

Referring to Figure 2.3, the plastic work done on a unit volume of material deformed in the tensile test to a true strain of ε_1 (where $\sigma_2 = \sigma_3 = 0$) will be, from Equation 2.15(b),

$$\frac{W}{\text{vol.}} = \int_0^{\varepsilon_1} \frac{dW}{\text{vol.}} = \int_0^{\varepsilon_1} \sigma_1 \, d\varepsilon_1 \tag{2.16}$$

i.e. the work done per unit volume is equal to the area under the true stress–strain curve, shown shaded in Figure 2.3.

2.7 Work hardening hypothesis

In Section 2.5 it was shown that at a particular instant in a plane stress process where the flow stress, σ_f, was known, the stresses and the ratio of the strain increments for a small deformation could be determined. To model a process we need to be able to follow the deformation along the given loading path as the flow stress changes. Clearly we would need to know the strain hardening characteristic of the material as determined, for example, by the true stress–strain curve in the tensile test.

It has been found by experiment that the flow stress increases in any process according to the amount of plastic work done during this process; i.e. in two different processes, if the work done in each is the same, the flow stress at the end of each process will be the same regardless of the stress path. This statement is only true for monotonic processes that follow the conditions given in Section 2.2.4; if there is a reversal in the process, the flow stress cannot be predicted by any simple theory.

In a plane stress, proportional process, we can plot the relation between each of the non-zero stresses and its strain as shown in Figure 2.11. From Equation 2.15(b), the sum of the shaded areas shown is the total work done per unit volume of material in the process. According to this *work-hardening hypothesis,* the flow stress at the end of this process is that given by the tensile test curve, Figure 2.3, when an equal amount of work has been done, i.e. when the sum of the areas in Figure 2.11 is equal to the area under the curve in Figure 2.3.

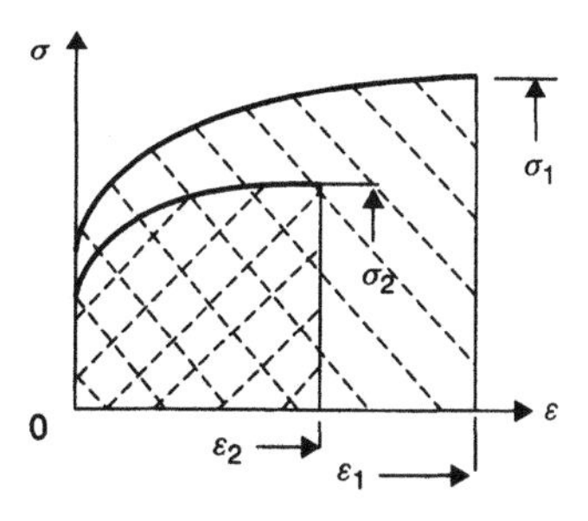

Figure 2.11 Stress–strain curves for the principal directions 1 and 2 for an element deforming in a plane stress procedure in which $\sigma_2 = \alpha\sigma_1$.

The way in which this work-hardening hypothesis is implemented in any analysis is described in the next section.

2.8 Effective stress and strain functions

The plastic work done per unit volume in an increment in a process is given by Equation 2.15(a). It would useful if this could be expressed in the form

$$\frac{\mathrm{d}W}{\text{vol.}} = f_1(\sigma_1, \sigma_2, \sigma_3)\mathrm{d}f_2(\varepsilon_1, \varepsilon_2, \varepsilon_3) \tag{2.17}$$

As the element is yielding during deformation, a suitable stress function to choose is that given by the von Mises yielding criterion, which has already been shown to have the value of the flow stress. For plane stress this function is,

$$f_1(\sigma_1, \sigma_2, \sigma_3) = \left(\sqrt{1 - \alpha + \alpha^2}\right)\sigma_1$$

This function is called the *representative, effective* or *equivalent stress*, $\bar{\sigma}$, and if the material is yielding, it will be equal to the flow stress. For a general state of stress in an isotropic material the effective stress function is, from Equation 2.12(b),

$$\bar{\sigma} = \sqrt{\frac{1}{2}\{(\sigma_1 - \sigma_2)^2 + (\sigma_2 - \sigma_3)^2 + (\sigma_3 - \sigma_1)^2\}} \tag{2.18a}$$

In plane stress, the effective stress function is

$$\bar{\sigma} = \sqrt{\sigma_1^2 - \sigma_1\sigma_2 + \sigma_2^2} = \left(\sqrt{1 - \alpha + \alpha^2}\right)\sigma_1 \tag{2.18b}$$

As indicated, if the material element is at yield, this function will have the magnitude of the flow stress, σ_f.

The required strain function in Equation 2.17 can be found by substitution of the stress function. This function is known as the *representative, effective* or *equivalent strain increment* $\mathrm{d}\bar{\varepsilon}$ and for plane stress, the function is

$$\mathrm{d}\bar{\varepsilon} = \mathrm{d}f_2(\varepsilon_1, \varepsilon_2, \varepsilon_3) = \sqrt{\frac{4}{3}\{1 + \beta + \beta^2\}}\mathrm{d}\varepsilon_1 \tag{2.19a}$$

In a general state of stress it can be written as

$$\begin{aligned} \mathrm{d}\bar{\varepsilon} &= \sqrt{\frac{2}{3}\{\mathrm{d}\varepsilon_1^2 + \mathrm{d}\varepsilon_2^2 + \mathrm{d}\varepsilon_3^2\}} \\ &= \sqrt{\frac{2}{9}\{(\mathrm{d}\varepsilon_1 - \mathrm{d}\varepsilon_2)^2 + (\mathrm{d}\varepsilon_2 - \mathrm{d}\varepsilon_3)^2 + (\mathrm{d}\varepsilon_3 - \mathrm{d}\varepsilon_1)^2\}} \end{aligned} \tag{2.19b}$$

In a monotonic, proportional process, Equations 2.19(a) and (b) can be written in the integrated form with the natural or true strains ε substituted for the incremental strains $\mathrm{d}\varepsilon$; i.e.

$$\begin{aligned} \bar{\varepsilon} &= \sqrt{\frac{4}{3}\{1 + \beta + \beta^2\}}\varepsilon_1 \\ &= \sqrt{\frac{2}{3}\{\varepsilon_1^2 + \varepsilon_2^2 + \varepsilon_3^2\}} \\ &= \sqrt{\frac{2}{9}\{(\varepsilon_1 - \varepsilon_2)^2 + (\varepsilon_2 - \varepsilon_3)^2 + (\varepsilon_3 - \varepsilon_1)^2\}} \end{aligned} \tag{2.19c}$$

where, $\bar{\varepsilon}$, is the *representative, effective,* or *equivalent strain.*

Because of the way in which these relations have been derived, it can be seen that the work done per unit volume in any process is given by

$$\frac{W}{\text{vol.}} = \int_0^{\overline{\varepsilon}} \overline{\sigma}\,\overline{\varepsilon} \tag{2.20}$$

It is also evident that because the stress function has been chosen as the von Mises stress which is equal in magnitude to the flow stress when the material is deforming, the effective strain function will be equal to the strain in uniaxial tension when equal amounts of work are done in the general process and in uniaxial tension. Thus we have identified a general stress–strain relation for an isotropic material deforming plastically, namely the *effective stress–strain curve*, $\overline{\sigma} = f(\overline{\varepsilon})$; this is coincident with the tensile test true stress–strain curve for an isotropic material.

The key to this principle of the equivalence of plastic work done is illustrated in Figure 2.11. To reiterate, in a plane stress process, there are two stress strain curves. These must continuously satisfy the yield criterion and the condition that the work done in the process, i.e. the area under both curves, is equal to the work done in uniaxial tension. This work done determines the current yield stress in uniaxial tension which is also the flow stress. The effective stress and strain functions ensure that these conditions are met and enable the current flow stress for a material element deformed in any process to be determined from an experimental stress–strain curve obtained in a tension test. Material properties can also be obtained from other tests, provided that the test enables an effective stress–strain curve to be obtained.

2.9 Summary

In this chapter, it is shown that for simple (monotonic, proportional), plane stress processes, it is possible to determine at any instant the principal membrane stresses required for deformation provided the current flow stress σ_f is known and also that either the stress ratio α or strain ratio β are known. The current flow stress can be determined from the tensile test stress strain curve using the effective stress and strain functions that are based on the equivalence of work. In practice, a process is often defined by the strain ratio β obtained from measurement of final strains. The assumption is that this point is reached by a proportional process, but if only the initial and final conditions are known, care should be taken in assuming that the strain ratio is constant.

The theory given in this chapter applies only for an instantaneous state in which the strain increment is small and the flow stress constant. In the next chapter, entire loading paths are studied using the theory established here. It cannot be emphasized too strongly that while the theory of deformation given here is useful and practical, it is a simple approximation of a very complex process. It is useful in process design and failure diagnosis in industry, but in some studies, more elaborate theories may be necessary.

2.10 Exercises

Ex. 2.1 A square element 8 × 8 mm in an undeformed sheet of 0.8 mm thickness becomes a rectangle, 6.5 × 9.4 mm after forming. Assume that the stress strain law is:

$$\overline{\sigma} = 600(0.008 + \overline{\varepsilon})^{0.22}\text{(MPa)}$$

and that the stress normal to the sheet is zero. Determine:

(a) the final membrane stresses;
(b) the final thickness;
(c) the principal strains.

Sketch these in the stress or strain space.
Also determine:

(d) the stress and strain ratios (assumed constant) and the hydrostatic stress and the deviatoric stresses at the end of the process, and
(e) the plastic work of deformation in the element.

[Ans: (a) 151.3 MPa, −336.6 MPa; (b) 0.838 mm; (c) 0.161, −0.208; (d) −2.29, −1.29, −61.7 MPa, 213 MPa, −274.9 MPa, 61.7 MPa; (e) 4.01 J]

Ex. 2.2 For a stress state in which the intermediate stress is $\sigma_2 = 0.5\ (\sigma_1 + \sigma_3)$, show that in yielding with a von Mises criterion and flow stress, σ_f, the diameter of the Mohr circle of stress is $(2/\sqrt{3})\sigma_f$.

Ex. 2.3 Show that in uniaxial deformation in which $\sigma_1 = \sigma_f$, $\sigma_2 = \sigma_3 = 0$, that the effective stress and effective strain increment are:

$$\bar{\sigma} = \sigma_1 = \sigma_f$$

$$d\bar{\varepsilon} = d\varepsilon_1$$

Ex. 2.4 Three principal stresses are applied to a solid where $\sigma_1 = 400\,\text{MPa}$, $\sigma_2 = 200\,\text{MPa}$, and $\sigma_3 = 0$.

(a) What is the ratio $d\varepsilon_1/d\varepsilon_3$?
(b) If a fluid pressure produces an all-around hydrostatic stress of −250 MPa that is superimposed upon the original stress state, how does the ratio in (a) change ? Explain the result.

[Ans: (a) −1; (b) no change.]

Ex. 2.5 Compare two plane stress deformation processes for a sample of high-strength low alloy steel of 1.2 mm thickness. In (a), biaxial tension, the strain ratio is 1 and in the other, (b), shear or drawing, the strain ratio is −1. A square element whose sides are aligned with the principal directions is initially 10 by 10 mm. In each case, one side is extended to 12 mm. The material properties are described by,

$$\bar{\sigma} = 850\bar{\varepsilon}^{0.16}\ \text{MPa}$$

Compare the final effective strains and stresses, the principal strains and stresses and the final sheet thickness.

[Ans:	$\bar{\varepsilon}$	$\bar{\sigma}$	$\varepsilon_1, \varepsilon_2$	σ_1, σ_2	t
(a)	*0.365,*	*723 MPa,*	*0.182, 0.182,*	*723, 723 Mpa,*	*0.833 mm.*
(b)	*0.211*	*662 MPa*	*0.182, −0.182,*	*382, −382 MPa*	*1.2 mm]*

Ex. 2.6 Assuming a von Mises criterion, a flow stress of σ_f and incompressbility, complete the table given below:

Stress/strain state	σ_1	σ_2	σ_3	$d\varepsilon_1$	$d\varepsilon_2$	$d\varepsilon_3$
Plane stress Pure shear 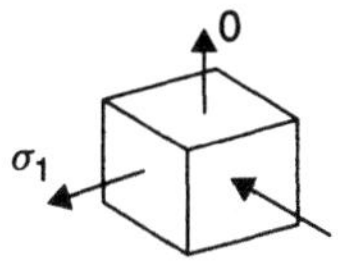			0	$d\varepsilon_1$		
Plane stress plane strain 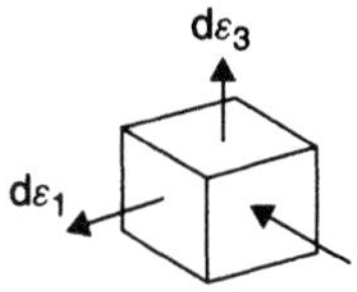			0	$d\varepsilon_1$	0	
Biaxial tension 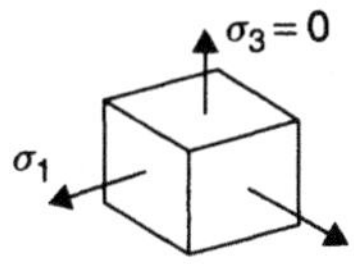			0	$d\varepsilon_1$		

3 Deformation of sheet in plane stress

3.1 Uniform sheet deformation processes

In Chapter 2, an instant in the plane stress deformation of a work-hardening material was considered. We now apply the theory to some region of a sheet undergoing uniform, proportional deformation as shown in Figure 3.1. If the undeformed sheet, of initial thickness t_0 is marked with a grid of circles of diameter d_0 or a square mesh of pitch d_0 as shown in Figure 3.1(a), then during uniform deformation, the circles will deform to ellipses of major and minor axes d_1 and d_2 respectively. If the square grid is aligned with the principal directions, it will become rectangular as shown in Figure 3.1(b). The thickness is denoted by t. At the instant shown in Figure 3.1(b), the deformation stresses are σ_1 and σ_2.

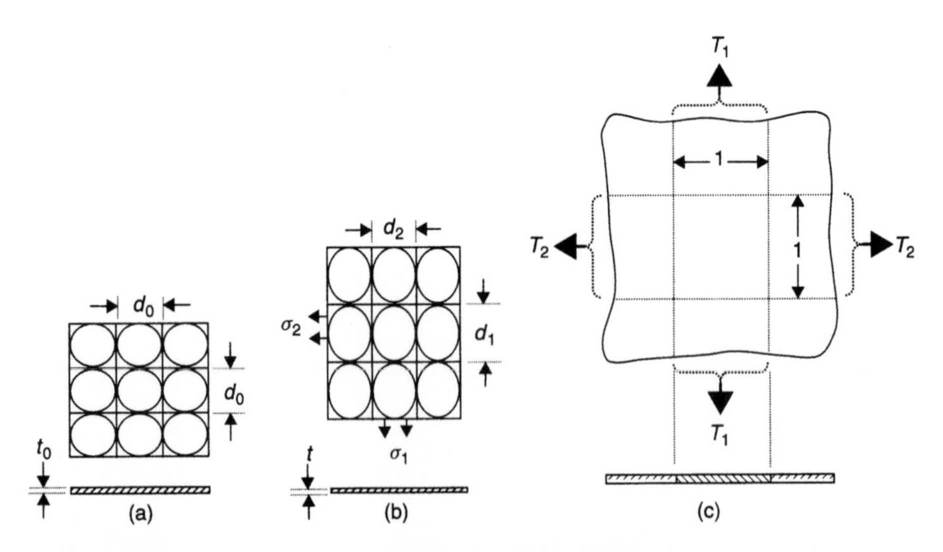

Figure 3.1 An element of a sheet showing: (a) the undeformed state with circle and square grids marked on it; (b) the deformed state with the grid circles deformed to ellipses of major diameter d_1 and minor diameter d_2 and (c) the tractions, T, or forces transmitted per unit width.

3.1.1 Tension as a measure of force in sheet forming

In sheet processes, deformation occurs as the result of forces transmitted through the sheet. The force per unit width of sheet is the product of stress and thickness and in Figure 3.1(c) is represented by,

$$T = \sigma t$$

where, T, is known as the *tension, traction* or *stress resultant.* Because this is the product of the current thickness t as well as the current stress σ, it is the appropriate measure of force and will be used throughout this work in modelling processes. The term, *tension,* will be used even though this suffers from the disadvantage that the force is not always a tensile force. If the tension is negative, it indicates a compressive force. This is not a serious problem as in plane stress sheet forming, almost without exception, one tension will be positive, i.e. the sheet is always pulled in one direction. It is impractical to form sheet by pushing on the edge; the expression used by practical sheet formers is that 'you cannot push on the end of a rope'.

In the convention used here, the principal direction 1 is that in which the principal stress has the greatest (most positive) value, and the major tension $T_1 = \sigma_1 t$ will always be positive. In stretching processes, the minor tension $T_2 = \sigma_2 t$ is tensile or positive. In other processes, the minor tension could be compressive and in some cases the thickness will increase. If T_2 is compressive and large in magnitude, wrinkling may be a problem.

In discussing true stress in Section 1.1.3, it was shown that for most real materials, strain-hardening continues, although at a diminishing rate, and true stress does not reach a maximum. As tension includes thickness, which in many processes will diminish, T may reach a maximum; this limits the sheet's ability to transmit load and is one of the reasons for considering tension in any analysis.

3.2 Strain distributions

In the study of any process, we usually determine first the strain over the part. This can be done by measuring a grid as in Figure 3.1, or by analysis of the geometric constraint exerted on the part. An example is the deep drawing process in Figure 3.2(a) and in the Introduction, Figure I.9. As the process is symmetric about the axis, we need only consider the strain at points on a line as shown in Figure 3.2(b). Plotting these strains in the principal strain space, Figure 3.2(c), gives the locus of strains for a particular stage in the process. As the process continues, this locus will expand, but not necessarily uniformly; some points may stop straining, while others go on to reach a process limit.

For any process, there will be a characteristic strain pattern, as shown in Figure 3.2(c). This is sometimes known as the 'strain signature'. Considerable information can be obtained from such a diagram and the way it is analysed is outlined in the following section.

3.3 Strain diagram

The individual points on the strain locus in Figure 3.2(c) can be obtained from measurements of a grid circles as shown in Figure 3.1. (If a square grid is used, the analysis

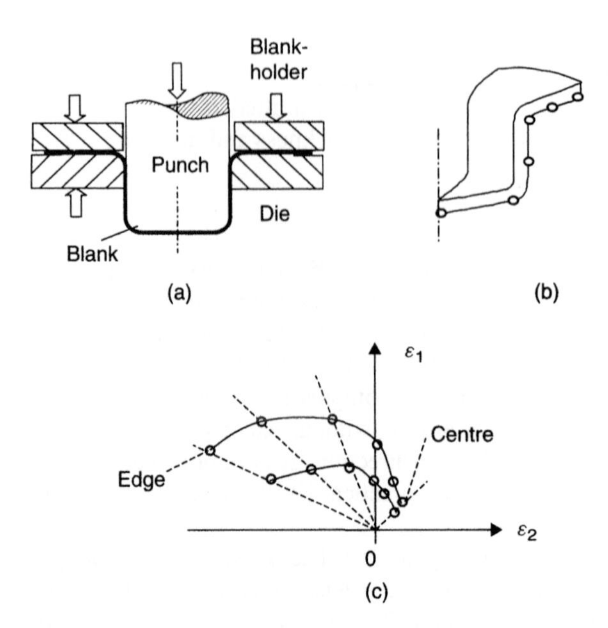

Figure 3.2 (a) Deep drawing a cylindrical cup. (b) Sector of a cup showing the location of strain measurements. (c) Strain plots for two stages in the drawing process.

method is outlined in Appendix A.2.) If the major and minor axes are measured and the current thickness determined, the analysis is as follows.

3.3.1 Principal strains

The principal strains at the end of the process are

$$\varepsilon_1 = \ln \frac{d_1}{d_0}; \quad \varepsilon_2 = \ln \frac{d_2}{d_0}; \quad \varepsilon_3 = \ln \frac{t}{t_0} \tag{3.1a}$$

3.3.2 Strain ratio

It is usual to assume that the strain path is linear, i.e. the strain ratio remains constant and is given by

$$\beta = \frac{\varepsilon_2}{\varepsilon_1} = \frac{\ln (d_2/d_0)}{\ln (d_1/d_0)} \tag{3.2}$$

In practice, care must be taken to determine if this assumption is reasonable. There are cases in which the strain path will deviate significantly from linearity. Such cases cannot be analysed in any simple way.

3.3.3 Thickness strain and thickness

From Equation 3.1(a), the thickness strain is determined by measurement of thickness, or alternatively from the major and minor strains assuming constant volume deformation, i.e.

$$\varepsilon_3 = \ln \frac{t}{t_0} = -(1+\beta)\,\varepsilon_1 = -(1+\beta)\ln \frac{d_1}{d_0} \tag{3.3}$$

From Equation 3.3, the current thickness is

$$t = t_0 \exp(\varepsilon_3) = t_0 \exp[-(1+\beta)\varepsilon_1] \tag{3.4}$$

or alternatively, as the volume $td_1d_2 = t_0d_0^2$ remains constant,

$$t = t_0 \frac{d_0^2}{d_1 d_2} \tag{3.5}$$

3.3.4 Summary of the deformation at a point

From the above, the principal strains and the strain ratio can be determined. The straining process is conveniently described by the principal strains, i.e.

$$\varepsilon_1 = \ln\left(\frac{d_1}{d_0}\right); \quad \varepsilon_2 = \ln\left(\frac{d_2}{d_0}\right) = \beta\varepsilon_1; \quad \varepsilon_3 = \ln\left(\frac{t}{t_0}\right) = -(1+\beta)\varepsilon_1 \tag{3.1b}$$

where β is constant.

Each point in the strain diagram in Figure 3.2(c) indicates the magnitude of the final major and minor strain and the assumed linear path to reach this point. Referring to Figure 3.3(a), we examine in more detail the character of different strain paths. This diagram, Figure 3.3(a) does not represent any particular process, but will be used to discuss the different deformation processes. The ellipse shown is a contour of equal effective strain, $\bar{\varepsilon}$; each point on the ellipse will represent strain in a material element that, from the work-hardening hypothesis in Section 2.7, has the same flow stress, σ_f.

3.4 Modes of deformation

If, by convention, we assign the major principal direction 1 to the direction of the greatest (most positive) principal stress and consequently greatest principal strain, then all points will be to the left of the right-hand diagonal in Figure 3.3(a), i.e. left of the strain path in

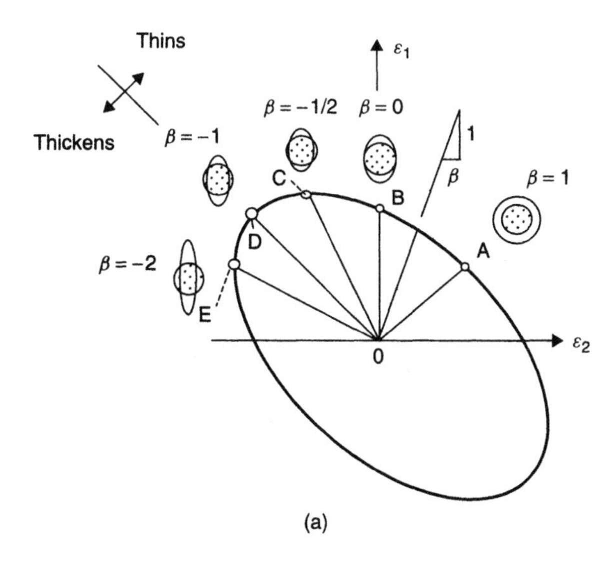

(a)

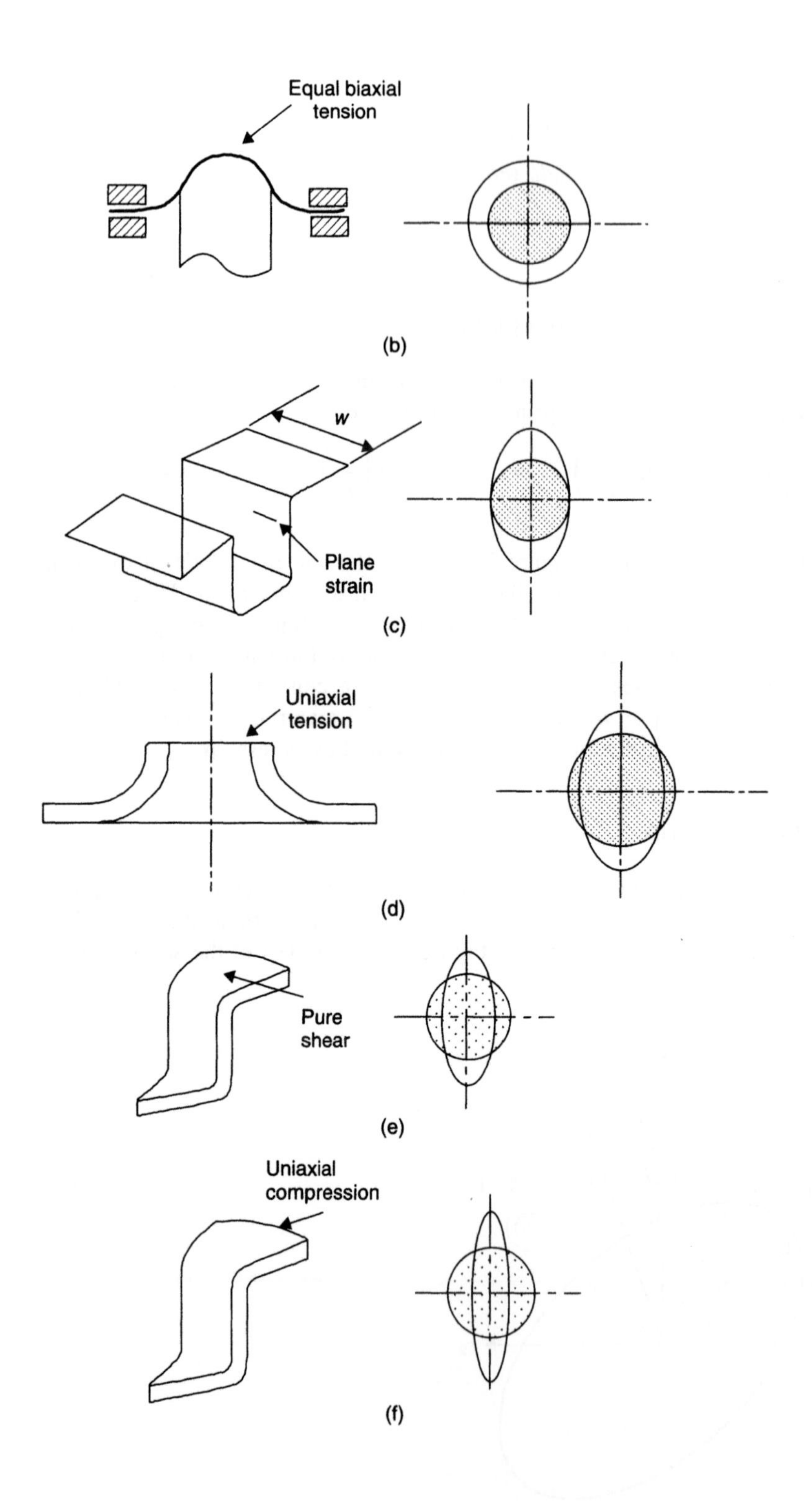

Equal biaxial tension
(b)
w
Plane strain
(c)
Uniaxial tension
(d)
Pure shear
(e)
Uniaxial compression
(f)

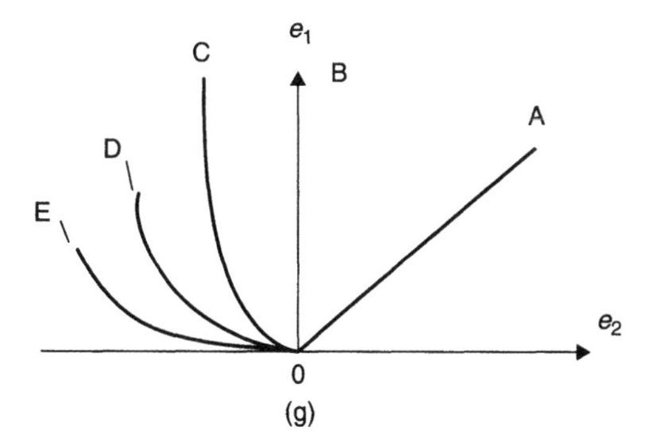

Figure 3.3 (a) The strain diagram showing the different deformation modes corresponding to different strain ratios. (b) Equibiaxial stretching at the pole of a stretched dome. (c) Deformation in plane strain in the side-wall of a long part. (d) Uniaxial extension of the edge of an extruded hole. (e) Drawing or pure shear in the flange of a deep-drawn cup, showing a grid circle expanding in one direction and contracting in the other. (f) Uniaxial compression at the edge of a deep-drawn cup. (g) The different proportional strain paths shown in Figure 3.2 plotted in an engineering strain diagram.

which $\beta = 1$. As stated above, the principal tension and principal stress in the direction, 1, will always be tensile or positive, i.e. $\sigma_1 \geq 0$. For the extreme case in which $\sigma_1 = 0$ we find from Equations 2.6 and 2.14, that $\alpha = -\infty$ and $\beta = -2$. Therefore all possible straining paths in sheet forming processes will lie between 0A and 0E in Figure 3.3(a) and the strain ratio will be in the range $-2 \leq \beta \leq 1$.

3.4.1 Equal biaxial stretching, $\beta = 1$

The path 0A indicates equal biaxial stretching. Sheet stretched over a hemispherical punch will deform in this way at the centre of the process shown in Figure 3.3(b). The membrane strains are equal in all directions and a grid circle expands, but remains circular. As $\beta = 1$, the thickness strain is $\varepsilon_3 = -2\varepsilon_1$, so that the thickness decreases more rapidly with respect to ε_1 than in any other process. Also from Equation 2.19(c), the effective strain is $\overline{\varepsilon} = 2\varepsilon_1$ and the sheet work-hardens rapidly with respect to ε_1.

3.4.2 Plane strain, $\beta = 0$

In this process illustrated by path, 0B, in Figure 3.3(a), the sheet extends only in one direction and a circle becomes an ellipse in which the minor axis is unchanged. In long, trough-like parts, plane strain is observed in the sides as shown in Figure 3.3(c). It will be shown later that in plane strain, sheet is particularly liable to failure by splitting.

3.4.3 Uniaxial tension, $\beta = -1/2$

The point C in Figure 3.3(a) is the process in a tensile test and occurs in sheet when the minor stress is zero, i.e. when $\sigma_2 = 0$. The sheet stretches in one direction and contracts in the other. This process will occur whenever a free edge is stretched as in the case of hole extrusion in Figure 3.3(d).

3.4.4 Constant thickness or drawing, $\beta = -1$

In this process, point D, membrane stresses and strains are equal and opposite and the sheet deforms without change in thickness. It is called *drawing* as it is observed when sheet is drawn into a converging region. The process is also called pure shear and occurs in the flange of a deep-drawn cup as shown in Figure 3.3(e). From Equation 3.1(b), the thickness strain is zero and from Equation 2.19(c) the effective strain is $\overline{\varepsilon} = \left(2/\sqrt{3}\right)\varepsilon_1 = 1.155\varepsilon_1$ and work-hardening is gradual. Splitting is unlikely and in practical forming operations, large strains are often encountered in this mode.

3.4.5 Uniaxial compression, $\beta = -2$

This process, indicated by the point E, is an extreme case and occurs when the major stress σ_1 is zero, as in the edge of a deep-drawn cup, Figure 3.3(f). The minor stress is compressive, i.e. $\sigma_2 = -\sigma_f$ and the effective strain and stress are $\overline{\varepsilon} = -\varepsilon_2$ and $\overline{\sigma} = -\sigma_2$ respectively. In this process, the sheet thickens and wrinkling is likely.

3.4.6 Thinning and thickening

Plotting strains in this kind of diagram, Figure 3.3(a), is very useful in assessing sheet forming processes. Failure limits can be drawn also in such a space and this is described in a subsequent chapter. The position of a point in this diagram will also indicate how thickness is changing; if the point is to the right of the drawing line, i.e. if $\beta > -1$, the sheet will thin. For a point below the drawing line, i.e. $\beta < -1$, the sheet becomes thicker.

3.4.7 The engineering strain diagram

In the sheet metal industry, the information in Figure 3.3(a) is often plotted in terms of the engineering strain. In Figure 3.3(g), the strain paths for constant *true strain* ratio paths have been plotted in terms of engineering strain. It is seen that many of these proportional processes do not plot as straight lines. This is a consequence of the unsuitable nature of engineering strain as a measure of deformation and in this work, true strains will be used in most instances. Engineering strain diagrams are still widely used and it is advisable to be familiar with both forms. In this work, true strain diagrams will be used unless specifically stated.

3.5 Effective stress–strain laws

In the study of a process, the first step is usually to obtain some indication of the strain distribution, as in Figure 3.2(c). As mentioned, this may be done by measuring grids or from some geometric analysis. The next step is to determine the stress state associated with strain at each point. To do this, one must have stress–strain properties for the material and Chapter 2 indicates how the tensile test data can be generalized to apply to any simple process using the effective stress–strain relations. In numerical models, the actual stress–strain curve can be used as input, but in a mechanics model it is preferable to use a simple empirical law that approximates the data. Here we consider some of these laws.

The effective strain $\bar{\varepsilon}$ for any deformation process such as the one illustrated in Figure 3.1 can be calculated from the principal strains and the strain ratio using Equation 2.19(c). As shown in Section 2.8, if the material is isotropic, the effective stress–strain curve is coincident with the uniaxial true stress–strain curve and a variety of mathematical relations may be fitted to the true stress–strain data. Some of the more common empirical relations are shown in Figure 3.4 and in these diagrams elastic strains are neglected. In the diagrams shown, the experimental curve is represented by a light line, and the fitted curve by a bold line.

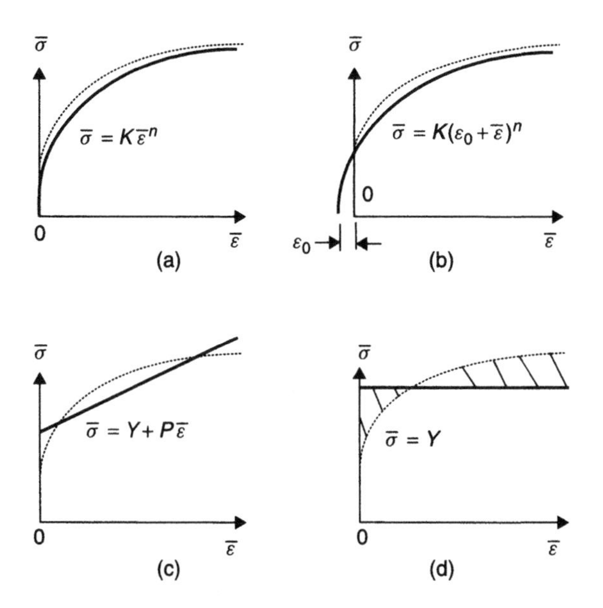

Figure 3.4 Empirical effective stress–strain laws fitted to an experimental curve.

3.5.1 Power law

A simple power law

$$\bar{\sigma} = K\bar{\varepsilon}^n \tag{3.6}$$

will fit data well for some annealed sheet, except near the initial yield; this is shown in Figure 3.4(a). The exponent, n, is the *strain-hardening index* as described in Section 1.1.3. The constants, K and n, are obtained by linear regression as explained in the section referred to. The only disadvantage of this law is that at zero strain, it predicts zero stress and an infinite slope to the curve. It does not indicate the actual initial yield stress.

3.5.2 Use of a pre-strain constant

Although it requires the determination of three constants, a law of the type

$$\bar{\sigma} = K\,(\varepsilon_0 + \bar{\varepsilon})^n \tag{3.7}$$

is useful and will fit a material with a definite yield stress as shown in Figure 3.4(b). The constant ε_0 has been termed a *pre-strain* or *offset strain* constant. If the material has been hardened in some prior process, this constant indicates a shift in the strain axis corresponding to this amount of strain as shown in Figure 3.4(b). In materials which are very nearly fully annealed and for which ε_0 is small, this relation can be obtained by first fitting Equation 3.6 and then, using the same values of K and n, to determine the value of ε_0 by fitting the curve to the experimentally determined initial yield stress using the equation

$$(\sigma_f)_0 = K\varepsilon_0^n \tag{3.8}$$

3.5.3 Linear strain-hardening

Although the fit is not good, a relation of the form

$$\overline{\sigma} = Y + P\overline{\varepsilon} \tag{3.9}$$

may sometimes be used, as shown in Figure 3.4(c). This law is appropriate for small ranges of strain.

3.5.4 Constant flow stress (rigid, perfectly plastic model)

In approximate models, strain-hardening may be neglected and the law

$$\overline{\sigma} = Y \tag{3.10}$$

employed. If the strain range of the process is known, the value of Y may be chosen so that the work calculated from the law will equal the work done in the actual process, i.e. the area under the approximate curve will equal that under the real curve and the areas shaded in Figure 3.4(d) will be equal.

3.5.5 (Worked example) empirical laws

An aluminium alloy is used in a process in which the effective strain is $\overline{\varepsilon} = 0.40$. The stress–strain curve is fitted by a law $\overline{\sigma} = 500\overline{\varepsilon}^{0.26}$ MPa. (a) Determine the effective stress at the end of the process. (b) Determine a suitable value for stress in a constant flow stress law. (c) If the initial yield stress is 100 MPa, determine a suitable value for the constant ε_0, in the law, $\overline{\sigma} = K(\varepsilon_0 + \overline{\varepsilon})^n$ MPa. Compare the stress given by this law and the simple power law at the final strain.

Solution

(a) At $\overline{\varepsilon} = 0.40$, $\overline{\sigma} = 500 \times 0.4^{0.26} = 394$ MPa.

(b) For equal work done,

$$\overline{Y}\overline{\varepsilon} = \int_0^{\overline{\varepsilon}} K\overline{\varepsilon}^n \mathrm{d}\overline{\varepsilon} = \frac{K}{1+n}\left[\overline{\varepsilon}^{1+n}\right]_0^{\overline{\varepsilon}}$$

$$\therefore \overline{Y} = \frac{K\overline{\varepsilon}^n}{1+n} = \frac{394}{1.26} = 313\,\text{MPa}$$

(c) By substitution, at ε_0, $\overline{\sigma} = 500\varepsilon_0^{0.26} = 100\,\text{MPa} \therefore \varepsilon_0 = 0.20^{\frac{1}{0.26}} = 0.002$.

At $\overline{\varepsilon} = 0.40$, $\overline{\sigma} = K(\varepsilon_0 + \overline{\varepsilon}\,)^n$ gives $\overline{\sigma} = 395\,\text{MPa}$; i.e. the difference in the laws is negligible at large strains.

3.6 The stress diagram

It has already been mentioned that a diagram in which the strains are plotted, e.g. Figure 3.2(c), is valuable in the study of a process. In the same way, a diagram in which the stress state associated with each strain point is shown is very useful in understanding the forces involved in a process. Such a diagram is shown in Figure 3.5. Like Figure 3.3(a), this is not a diagram for a particular process, but is used to illustrate the link between the strain and stress diagrams. Also, contours of equal effective stress are shown, which are of course yield loci for particular values of flow stress. During deformation, plastic flow will start from the initial yield locus shown as a continuous line, i.e. when $\overline{\sigma} = (\sigma_f)_0$ and the loading path will be along a radial line of slope $1/\alpha$.

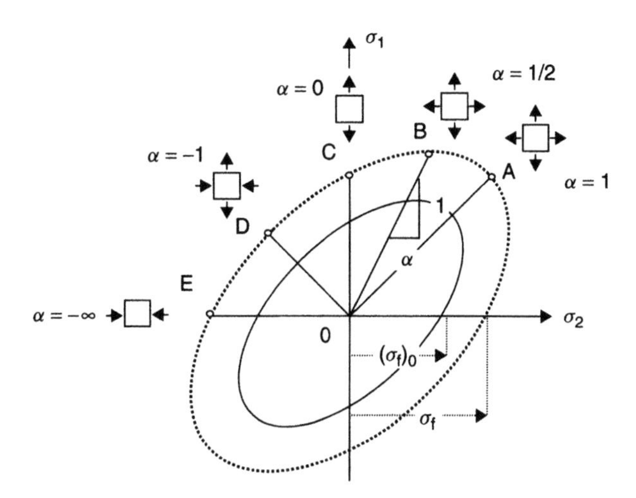

Figure 3.5 The processes shown in the strain space, Figure 3.2, illustrated here in the stress space. (The current yield ellipse is shown as a broken line.).

To plot a point in this diagram, the stress ratio is calculated from the strain ratio, Equation 2.14. The effective strain is determined from Equation 2.19(c), and from the known material law, the effective stress determined and the principal stresses calculated from Equation 2.18(b). The current state of stress is shown as a point on the ellipse given as a broken line. This yield locus intercepts the axes at $\pm\sigma_f$.

The principal stresses are

$$\sigma_1; \qquad \sigma_2 = \alpha.\sigma_1 \qquad \text{and } \sigma_3 = 0 \tag{3.11}$$

and each path in the strain diagram, Figure 3.3(a) has a corresponding path in the stress diagram as detailed below.

3.6.1 Equal biaxial stretching, $\alpha = \beta = 1$

At A, the sheet is stretching in equal biaxial tension and

$$\sigma_1 = \sigma_2 = \overline{\sigma} \tag{3.12}$$

In an isotropic material, each stress is equal to that in a simple tension test.

3.6.2 Plane strain, $\alpha = 1/2, \beta = 0$

For plane strain, i.e., zero strain in the 2 direction, the stress state is indicated by the point B and

$$\sigma_1 = \frac{2}{\sqrt{3}}\overline{\sigma} = 1.15\overline{\sigma} \text{ and } \sigma_2 = \frac{1}{2}\sigma_1 \tag{3.13}$$

For a material of given flow stress, the magnitude of the major stress, σ_1, is greater in this process than in any other.

3.6.3 Uniaxial tension, $\alpha = 0, \beta = -1/2$

This point is illustrated by C in Figure 3.5; the major stress is equal to the flow stress σ_f and the minor stress is zero. The process occurs in the tensile test, and as mentioned, at a free edge.

3.6.4 Drawing, shear or constant thickness forming, $\alpha = -1, \beta = -1$

Along the left-hand diagonal at D, the membrane stresses and strains are equal and opposite and there is no change in thickness. The stresses are

$$\sigma_1 = \frac{1}{\sqrt{3}}\sigma_f = 0.58\sigma_f = 0.58\overline{\sigma}$$

and

$$\sigma_2 = -\frac{1}{\sqrt{3}}\sigma_f = -0.58\sigma_f = -0.58\overline{\sigma} \tag{3.14}$$

It will be noted that the magnitudes of stresses to cause deformation are at a minimum in this process, i.e. in magnitude, they are only 58% of the stress required to yield a similar element in simple tension. This can be considered an ideal mode of sheet deformation as the stresses are low, there is no thickness change, and, as will be shown later, failure by splitting is unlikely.

3.6.5 Uniaxial compression, $\alpha = -\infty, \beta = -2$

This mode mostly occurs at a free edge in drawing a sheet as the stress on the edge of the sheet is zero. The minor stress is equal to the compressive flow stress, i.e.

$$\sigma_1 = 0 \text{ and } \sigma_2 = -\sigma_f = -\overline{\sigma} \tag{3.15}$$

As indicated, high compressive stresses are often associated with wrinkling of the sheet.

3.7 Principal tensions or tractions

In Section 3.1.1, the use of 'tensions' in the analysis of sheet metal forming was introduced. The principal tensions on a sheet element are illustrated in Figure 3.1(c). *Tension* is the force per unit length transmitted in the sheet and has the units of [force]/[length]; a typical unit used is kN/m. These tensions govern deformation in the sheet and the forces acting on the tooling. It is found more convenient to model processes in terms of tension rather than stress and for this reason, the determination of tensions for the different processes illustrated in Figures 3.2(a) and 3.5 is described.

We shall be concerned here only with tensions in the principal directions. As thickness will change, we must calculate both the current principal stresses and thickness in order to determine the principal tensions. For any region, as in Figure 3.1, the principal stresses and current thickness can be determined using the relations given above. The tensions will be in proportion to the stresses, i.e.

$$T_1 = \sigma_1 t; \qquad T_2 = \sigma_2 t = \alpha T_1 \tag{3.16}$$

For a region of a sheet in which the thickness is uniform, i.e. t = constant, the principal tensions will satisfy a *tension yield locus* that is geometrically similar to the yield stress locus as given in Figures 2.6 and 2.7. If the material obeys a von Mises yield condition, the principal tensions in the sheet at yield will satisfy a *generalized* or *effective yielding tension* relation given by

$$\begin{aligned} \overline{T} = \overline{\sigma} t &= \sqrt{T_1^2 - T_1 T_2 + T_2^2} \\ &= \sqrt{1 - \alpha + \alpha^2}\, T_1 \end{aligned} \tag{3.17}$$

This is illustrated in Figure 3.6.

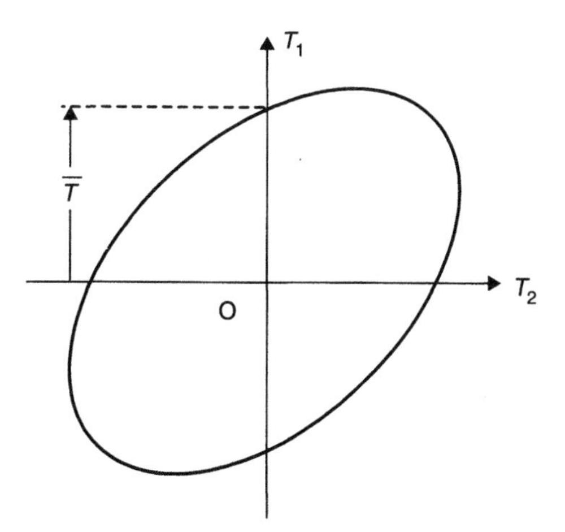

Figure 3.6 Relation between the principal tensions for an element deforming in a proportional process with a current effective tension of $\overline{T} = \overline{\sigma} t$.

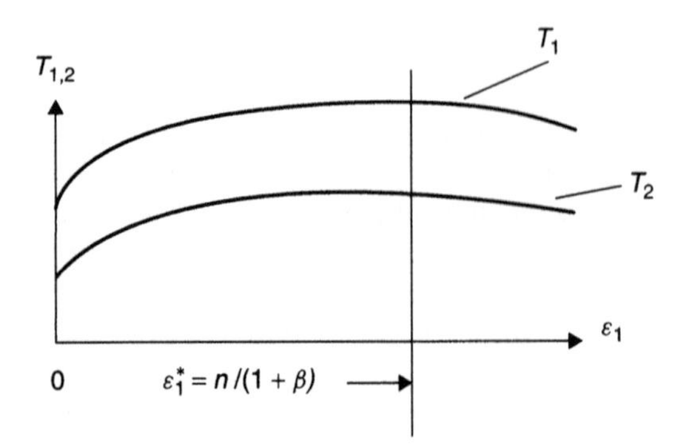

Figure 3.7 Principal tensions versus the major strain for a proportional process.

For any particular stress ratio and major strain, the effective stress and the thickness can be obtained using the relations given above. For a material in which the stresses and strains obey the power law, Equation 3.6, the major tension can be determined as

$$T_1 = \sigma_1 t = \frac{K\overline{\varepsilon}^n t_0 \exp\left[-(1+\beta)\,\varepsilon_1\right]}{\sqrt{1-\alpha+\alpha^2}} \tag{3.18}$$

This may be derived using equations 2.12(d), 2.5 and 2.6.
From Equations 3.16 and 3.18, the principal tensions can be found and are illustrated in Figure 3.7; in this case the strain ratio is positive. As discussed earlier, the major tension T_1 will always be equal to or greater than zero. Depending on the value of the stress or strain ratios, the minor tension T_2 will be positive or negative.

For a given material and process, Equation 3.18 can be reduced to the form

$$T_1 = const\ \varepsilon_1^n \exp\left[-(1+\beta)\,\varepsilon_1\right] \tag{3.19}$$

Differentiating this expression, we find that tensions reach a maximum only for processes in which the sheet thins, i.e. when $\beta > -1$. When this is the case, the strain at maximum tension, denoted by*, will be

$$\varepsilon_1^* = \frac{n}{1+\beta} \tag{3.20}$$

For uniaxial tension, $\beta = -1/2$, the maximum in tension is at $\varepsilon_1^* = 2n$, and for plane strain, $\beta = 0$, maximum tension is when $\varepsilon_1^* = n$. The relation between maximum tension and necking is discussed in Chapter 5.

3.7.1 (Worked example) tensions

Compare the tensions and the thickness at the point of maximum tension in a sheet initially of 0.8 mm thickness with a stress strain characteristic of, $\overline{\sigma} = 530(\overline{\varepsilon})^{0.246}$ MPa when deformed in equal biaxial tension and plane strain.

Solution. For equal biaxial tension, $\alpha = 1, \beta = 1$. At maximum tension,

$\varepsilon_1^* = \dfrac{n}{1+\beta} = \dfrac{n}{2} = 0.123$.The effective strain is $\overline{\varepsilon} = 2\varepsilon_1 = 0.246$.

The thickness at maximum tension is,

$t = t_0 \exp\left[-(1+\beta)\,\varepsilon_1^*\right] = 0.8 \exp(-0.246) = 0.626\,\text{mm}$

The effective stress is $\overline{\sigma} = 530 \times 0.246^{0.246} = 375\,\text{MPa}$.

The principal stresses are $\sigma_1 = \sigma_2 = \overline{\sigma}$ and

the principal tensions are $T_1 = T_2 = 375 \times 10^6 \times 0.626 \times 10^{-3} = 236\,\text{kN/m}$.

For plane strain, $\alpha = 0.5$, $\beta = 0$ and the strain at maximum tension is $\varepsilon_1^* = n = 0.246$ and $\overline{\varepsilon} = (2/\sqrt{3})\varepsilon_1$ and $\overline{\sigma} = (\sqrt{3}/2)\sigma_1$.

The thickness at maximum tension is,

$t = t_0 \exp\left[-(1+\beta)\,\varepsilon_1^*\right] = 0.8 \exp(-0.246) = 0.626\,\text{mm}$ and following the above,

$T_1 = 281\,\text{kN/m}, \qquad T_2 = \alpha T_1 = 140\,\text{kN/m}$

Thus for plane strain the maximum tension is greater, but the thickness at maximum tension is the same.

3.8 Summary

In the analysis of a sheet metal forming operation, different regions of the sheet will have a particular forming path that can often be considered as proportional and represented by a line in the stress and strain diagrams. These diagrams are the 'charts' in which the course of the process can be plotted. In subsequent chapters these diagrams will be used extensively.

3.9 Exercises

Ex. 3.1 A small circle (5.0 mm in diameter) is printed on the surface of an undeformed low carbon steel sheet with thickness 0.8 mm. Then the sheet is deformed in a plane stress proportional process. It is noted after unloading that the circle has been distorted into an ellipse with major and minor diameters of 6.1 mm and 4.8 mm respectively. The effective stress strain relation is:

$$\overline{\sigma} = 600\overline{\varepsilon}^{0.22}\,\text{MPa}$$

(a) Assuming that the loading is monotonic, what is the ratio of stresses α?
(b) Determine the tension T_1 and T_2.
(c) Calculate the effective strain.

[(Ans: 0.328; 329.2 kN/m, 108.2 kN/m; 0.21)]

Ex. 3.2 Consider the case of (a) constant thickness, (b) uniaxial tension, and (c) plane strain, where $\sigma_1 > \sigma_2$ and $\sigma_3 = 0$. For each case, compare the ratio of $\overline{\sigma}$, the effective stress, and τ_{max}, the maximum shear stress.
[Ans: $\sqrt{3}$; 2; $\sqrt{3}$).]

Ex. 3.3 In a deep-drawn cup as shown in Figure 3.2, the strains in the centre of the base, (a), half-way up the cup wall, (b), and in the middle of the flange,(c), are as follows: (a) 0.015, 0.015, (b) 0.050, 0.000, and (c) 0.150, −0.100. Strain-hardening in the material is negligible so that the effective stress is constant at 300 MPa. The initial thickness is

0.50 mm. Determine at each point, the thickness and the major tension (acting along the line shown in the sectioning plane).
[Ans: 0.485, 0.476, 0.476 mm; 146, 165, 125 kN/m]

Ex. 3.4 The average effective strain in a steel stamping is 0.03. If friction is neglected and all the plastic work done is converted into heat, what would be the average temperature rise? The effective stress strain law is $\overline{\sigma} = 600\,(0.01 + \overline{\varepsilon})^{0.22}$ MPa; for steel the specific heat is 454 J/kg/°C and the density is 7850 kg/m^3.
[Ans :≈ 2°C]

4 Simplified stamping analysis

4.1 Introduction

Stamping is the process of forming shallow parts in a press by stretching the sheet over a shaped punch and die set. In this chapter, stamping with a conventional *draw die* is considered; the cross-section of such a die is shown in Figure 4.1. In Figure 4.1(a), the blank is clamped at the edges by the *binder* or *blank-holder* using one action of the press. The binder holds the sheet in such a way that it can be drawn inwards against the clamping action but still develop sufficient tension to stretch the sheet over the punch, as shown in Figure 4.1(b). The resistance to drawing inwards is due to friction between the sheet and the binder and this may be enhanced by draw-beads that are described later. If some

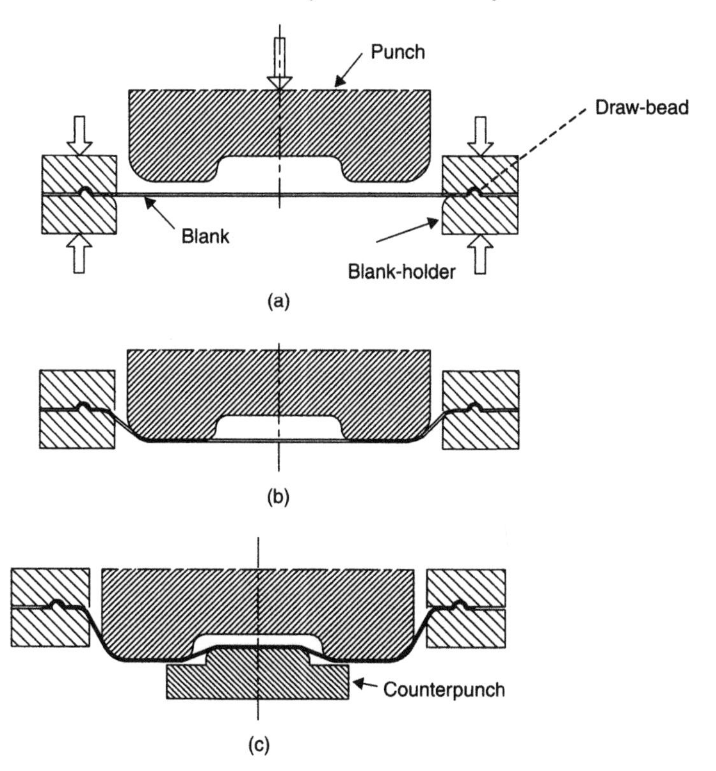

Figure 4.1 Cross-sectional view of a simple draw die.

re-entrant shape is required, this is formed when the punch bottoms on the lower die or *counterpunch* shown in Figure 4.1(c).

It is important to realize that even though matching dies may be used, the sheet is not compressed between them as in a forging process, but is stretched over each convex tool surface. Contact with the tooling is, for the most part, on only one side of the sheet; there is no through-thickness compression. In most regions the contact pressure is small compared with the flow stress of the sheet and it is usually acceptable to neglect through-thickness stresses and assume plane stress deformation.

Stamping is the basic process for forming parts whose shape cannot be obtained simply by bending or folding. For most autobody parts, the sheet is first formed to shape in a draw die in a *double-acting* press, i.e. one having separate clamping and punch actions as in Figure 4.1. Secondary forming and blanking operations may be carried out in subsequent presses. The shapes that are formed may be quite complex and the process is a three-dimensional one. Nevertheless, in this work we shall first consider very simple geometry as shown in Figure 4.2 and assume that this is a two-dimensional process, i.e. the strain perpendicular to the plane of the diagram is zero and the deformation is both plane stress and plane strain.

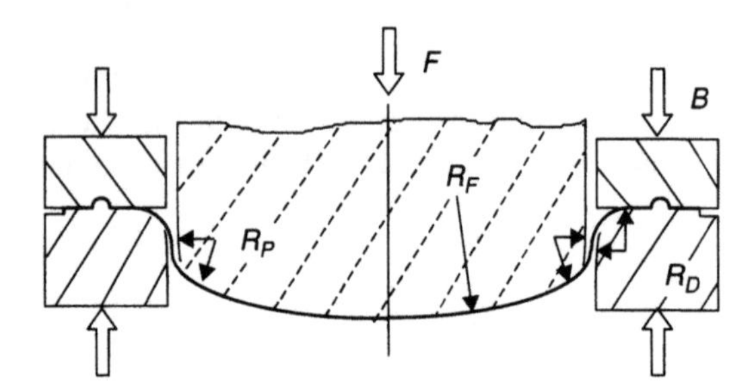

Figure 4.2 Simple draw die with the punch face having a circular profile.

4.2 Two-dimensional model of stamping

As the die in Figure 4.2 is symmetric, we consider only one half as shown in Figure 4.3. The punch has a cylindrical shape of radius R_F; the other features are defined in Figure 4.3 and are listed below:

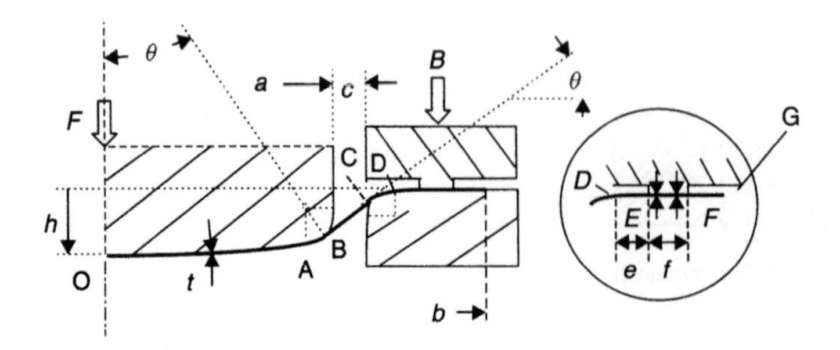

Figure 4.3 Half-section of a partially drawn strip in the die shown in Figure 4.2.

a	punch semi-width,
b	blank semi-width
c	side clearance
e	land width
f	width of frictional clamping (simulating a draw-bead)
h	punch penetration (part depth)
t	blank thickness
R_F	punch face radius
R_P	punch corner radius
R_D	die corner radius

(subscript, 0, denotes the initial value.)

Initially we do not consider draw-beads, but assume that the restraint is created only by friction acting over a specified area of the binder. At some depth of punch penetration, h, the different zones are shown in Figure 4.3 and listed below:

OB	material in contact with the punch,
BC	unsupported sheet in the side wall,
CD	sheet in contact with the die corner radius,
DE	sheet on the die land without contact pressure,
EF	region over which the blank-holder force acts,
FG	free edge of the blank.

We assume that all regions of the sheet from the centre-line O to the edge of the clamping area F are plastically deforming. (Often this will not be true as some regions may cease to deform even though the punch is still moving downwards.) From the centre O to the point of tangency B the sheet is stretching and sliding outwards against friction and the friction force on the sheet acts towards O. From the point of contact with the die C to the point F the sheet is sliding inwards and the friction force on the sheet acts outwards. If either the blank-holder force B or the strain at the centre is specified, it is possible to determine all other variables in a two-dimensional process.

4.2.1 Strain of an element

At any instant during stamping, the thickness and also the stress and tension will vary over the part. If we consider an infinitesimal element as shown in Figure 4.4, the conditions at a point will be as follows.

The principal direction, 1, is in the sectioning plane and the direction 2 is perpendicular to this. As the process is assumed to be one of plane strain, $\beta = 0$, the strain state is

$$\varepsilon_1; \qquad \varepsilon_2 = \beta\varepsilon_1 = 0; \qquad \varepsilon_3 = -(1+\beta)\varepsilon_1 = -\varepsilon_1 \tag{4.1}$$

The effective strain in the element is, from Equation 2.19(c),

$$\bar{\varepsilon} = \sqrt{\frac{4}{3}\left\{1+\beta+\beta^2\right\}}\varepsilon_1 = \frac{2}{\sqrt{3}}\varepsilon_1 \tag{4.2}$$

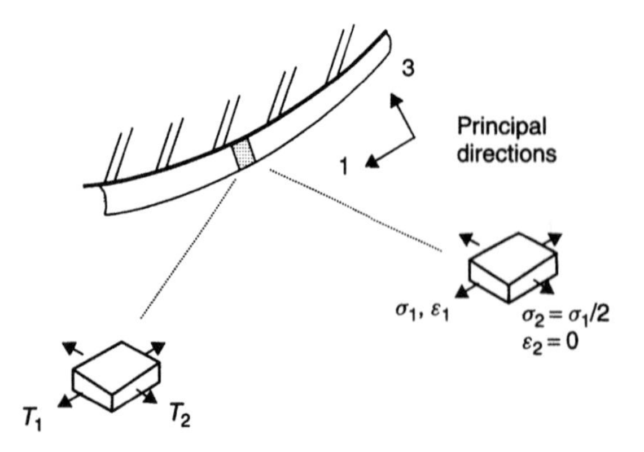

Figure 4.4 An element of the sheet sliding on the face of the punch.

4.2.2 Thickness of the element

The current thickness can be expressed in terms of the strain, ε_1, i.e.

$$t = t_0 \exp(\varepsilon_3) = t_0 \exp[-(1+\beta)\varepsilon_1] = t_0 \exp(-\varepsilon_1) \tag{4.3}$$

4.2.3 Stress on the element

The state of stress on the element is

$$\sigma_1; \qquad \sigma_2 = \alpha\sigma_1; \qquad \sigma_3 = 0 \tag{4.4}$$

and as $\beta = 0$, from Equation 2.14, $\alpha = 1/2$.

An effective stress–strain law must be chosen. In this case we shall choose one displaying a definite initial yield stress:

$$\overline{\sigma} = K(\varepsilon_0 + \overline{\varepsilon})^n \tag{3.7}$$

Combining this with Equations 4.2, we obtain

$$\overline{\sigma} = K\left[\varepsilon_0 + \left(2/\sqrt{3}\right)\varepsilon_1\right]^n \tag{4.5}$$

From this, the major stress σ_1 is obtained using Equation 2.18(b), i.e.

$$\sigma_1 = \overline{\sigma}/\sqrt{1-\alpha+\alpha^2} = 2\overline{\sigma}/\sqrt{3} \tag{4.6}$$

4.2.4 Tension or traction force at a point

As shown in Figure 3.7, for a given material and initial sheet thickness, the tension, or force per unit width at a point can be expressed as a function of the strain at that point. The major principal tension, T_1, in the sectioning plane is, from the above equations,

$$T_1 = \sigma_1 t = \frac{K\left[\varepsilon_0 + \sqrt{(4/3)\left(1+\beta+\beta^2\right)\varepsilon_1}\right]^n}{\sqrt{1-\alpha+\alpha^2}} t_0 \exp(-\varepsilon_1) \tag{4.7}$$

which, for the plane strain case in which $\beta = 0$ and $\alpha = 1/2$, gives

$$T_1 = \frac{2Kt_0}{\sqrt{3}} \left[\varepsilon_0 + \left(2/\sqrt{3}\right)\varepsilon_1\right]^n \exp\left(-\varepsilon_1\right) \tag{4.8}$$

and

$$T_2 = T_1/2 \tag{4.9}$$

Differentiating Equation 4.8, indicates that the tension reaches a maximum at a strain of

$$\varepsilon_1^* = n - \frac{\sqrt{3}}{2}\varepsilon_0 \tag{4.10}$$

where the 'star' denotes a limit strain for the process. Clearly when the tension reaches a maximum at a point, the sheet will continue to deform at that point under a falling tension. Other regions of the sheet will unload elastically and the sheet will fail at the point where the tension maximum occurred.

4.2.5 Equilibrium of the element sliding on a curved surface

We now consider a larger element of arc length ds as shown in Figure 4.5. If the tool surface is curved, there will be a contact pressure p and if the sheet is sliding along the surface, there will be a frictional shear stress μp, where μ is the coefficient of friction. Both the tension and the thickness will change because of the frictional force.

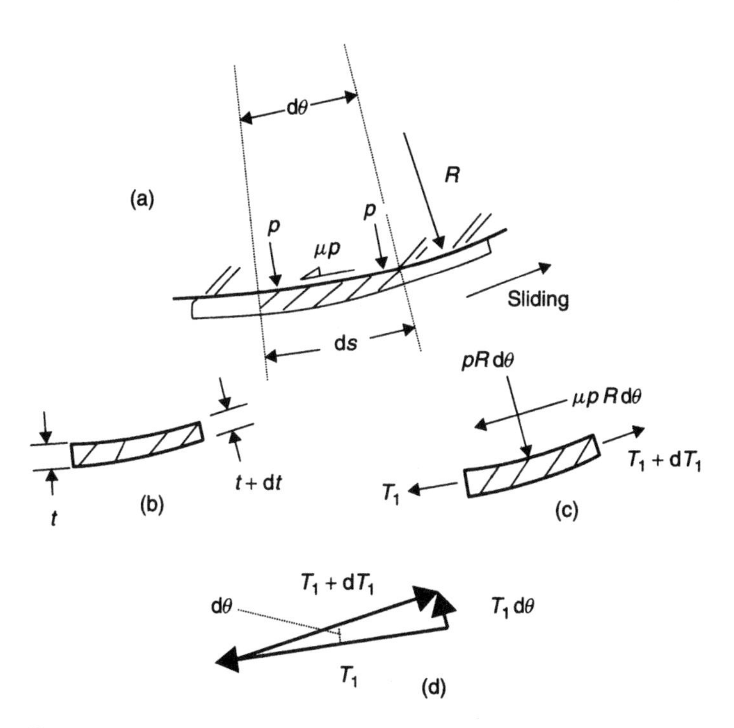

Figure 4.5 (a) An element sliding on a tool face. (b) Thickness of the element. (c) Forces on the element. (d) Resultant of tension forces acting radially inwards.

The length of the element can be expressed in terms of the tool radius and the angle subtended, i.e.

$$ds = R\,d\theta$$

and the surface area for a unit width of sheet is

$$R\,d\theta\,1$$

The force acting on the element radially outward is

$$pR\,d\theta$$

The force tangential to the sheet due to friction is

$$\mu pR\,d\theta$$

The tension forces are T_1 and $T_1 + dT_1$. As the direction of the tension forces differs by an angle, $d\theta$, there is a radially inward component of force, $T_1\,d\theta$, as shown in Figure 4.5(d).

The equilibrium equation for forces in the radial direction is

$$T_1\,d\theta = pR\,d\theta$$

or

$$p = \frac{T_1}{R} \tag{4.11}$$

It is useful at this stage to re-arrange Equation 4.11. Recalling that $T_1 = \sigma_1 t$, the contact pressure is

$$p = \frac{\sigma_1}{R/t} \tag{4.12}$$

The contact pressure as shown in Equation 4.12 is inversely proportional to the bend ratio R/t. The radius of curvature of the punch face is likely to be several orders of magnitude greater than the thickness and even at most corner radii, it will be 5 to 10 times the thickness. The principal stress σ_1 is at most only 15% greater that the flow stress σ_f, and so the contact pressure will be a small fraction of the flow stress, justifying the assumption of plane stress, except at very small radii in the tooling.

The equilibrium condition for forces along the sheet is, from Figure 4.5,

$$(T_1 + dT_1) - T_1 = \mu p 1 R\,d\theta$$

or, combining the above equations,

$$\frac{dT_1}{T_1} = \mu\,d\theta \tag{4.13a}$$

From Equation 4.12, the contact pressure depends on the radius ratio, but the change in tension as given in Equation 4.13(a) is independent of curvature and is a function of the coefficient of friction and the angle turned through, sometimes called the *angle of wrap*.

If the tension at one point, j, in the section is known, then the tension at some other point, k, can be found by integrating Equation 4.13(a), i.e.

$$\int_{T_{1j}}^{T_{1k}} \frac{dT_1}{T_1} = \int_0^{\theta_{jk}} \mu\,d\theta$$

or

$$T_{1k} = T_{1j} \exp \mu\theta_{jk} \tag{4.13b}$$

where, θ_{jk}, is the angle turned through between the two points. Care must be taken in using this relation to ensure that the material is sliding in the same direction everywhere between the two points and that there is no point of inflection in the surface profile.

4.2.6 Force equilibrium at the blank-holder and punch

At the region EF in Figure 4.3, the sheet is clamped between two flat surfaces by the blank-holder force. The force is expressed in terms of a *force per unit length*, B, as shown. A friction force, μB, acts on each side of the sheet and hence the equilibrium condition, as shown in Figure 4.6(a) is,

$$T_{1E} = 2\mu B \tag{4.14}$$

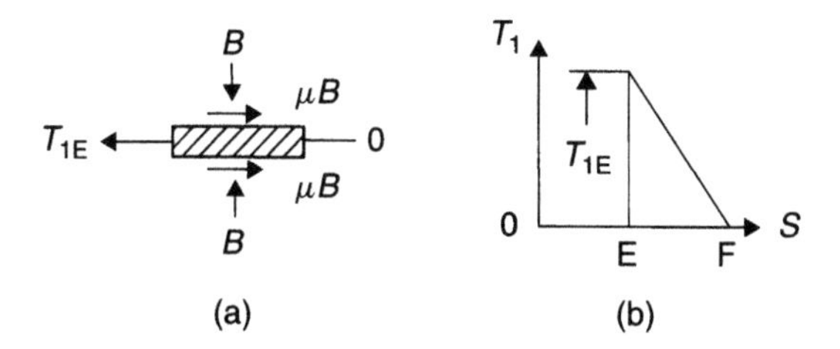

Figure 4.6 Equilibrium of the sheet under the blank-holder.

It is sufficiently accurate to assume that the tension will fall off linearly over the distance, EF, as shown in Figure 4.6(b).

As already mentioned, there is frequently a draw-bead instead of flat faces at a region such as EF. The effect will be similar in that the tension force will increase sharply over this region. The mechanics of draw-beads is given in a subsequent chapter, but in die design, a step change in tension (determined from experience) is often used to model the draw-bead action in an overall analysis of the process.

4.2.7 The punch force

The force acting on the punch, as shown in Figure 4.7, is in equilibrium with the tension in the side-wall. The angle of the side-wall θ_B can be obtained from the geometry. The vertical component of this tension force is $T_{1B} \sin\theta_B$ and therefore the punch force per unit width, considering both sides of the sheet, is

$$F = 2T_{1B} \sin\theta_B \tag{4.15}$$

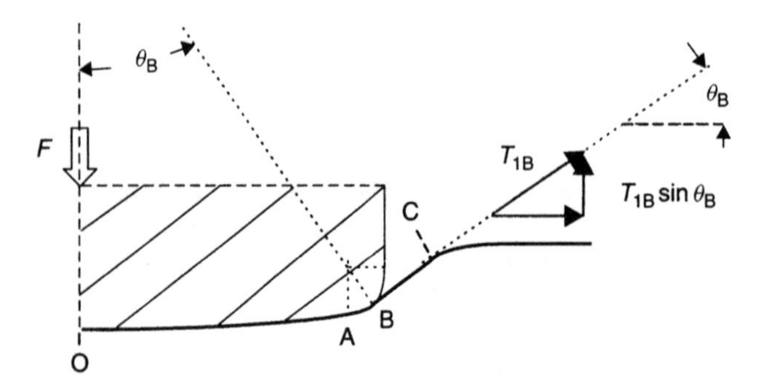

Figure 4.7 Diagram showing the relation between punch force and side-wall tension.

4.2.8 Tension distribution over the section

It is now possible to determine the tension at each point along a strip as illustrated in Figure 4.8. If the strain at the mid-point ε_{1O} is known or specified, the centre-line tension T_{1O} can be calculated from Equation 4.8. As the sheet between O and B is sliding outwards against an opposing friction force from B to O, the tension in the sheet will increase. The angle of wrap θ_B can be determined from the punch depth h and the tool geometry. The tension at B can be found from Equation 4.13(b), i.e.

$$T_{1B} = T_{1O} \exp(\mu.\theta_B)$$

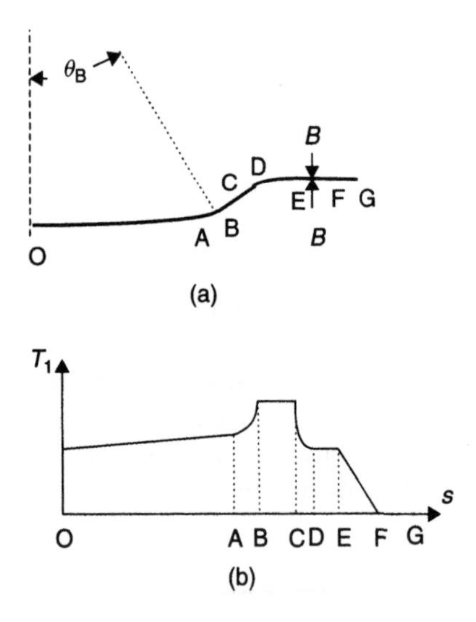

Figure 4.8 Distribution of tension forces across the sheet in a draw die.

In the side-wall, between B and C, the sheet is not in contact with the tooling and the tension is constant, i.e. $T_{1C} = T_{1B}$. If the surface of the sheet under the blank-holder is horizontal as shown, the angle turned through between C and D will be the same as θ_B and hence the tension at D, and also at E, will be equal to that at the centre-line, i.e. $T_{1D} = T_{1E} = T_{1O}$. From E to F, the tension falls to zero as indicated in Section 4.2.6.

The distribution of tension is shown in Figure 4.8(b). If the punch face is only gently curved, the angle of wrap and the tension will only increase slowly with distance along the strip from O to A. At the corner radius, the tension increases rapidly and reaches a plateau in the side-wall. It then drops down due to friction at the die corner radius and falls to zero outside the clamping area.

The blank-holder force required to generate this tension distribution is found from Equation 4.14. The higher the blank-holder force, the greater will be the strain over the punch face, however the process is limited by the strain in the side-wall. The tension here has a maximum value determined from Equations 4.8 and 4.10. If this maximum is reached, the side-wall will fail by splitting.

4.2.9 Strain and thickness distribution

The distribution of strain corresponding to the tension distribution in Figure 4.8 can be found from Equation 4.8. Writing this in the form

$$\left(\varepsilon_0 + (2/\sqrt{3})\varepsilon_1\right)^n \exp(-\varepsilon_1) = \sqrt{3}T_1/2Kt_0 \tag{4.16}$$

shows that ε_1 must be found by a numerical solution.

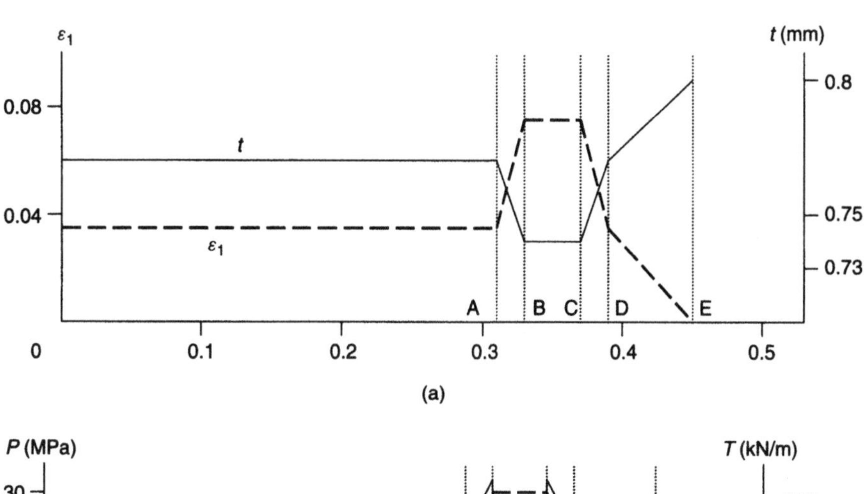

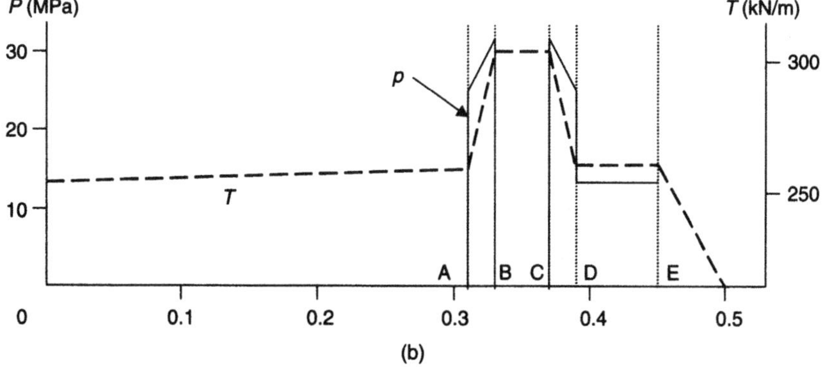

Figure 4.9 Distribution of strain, thickness, pressure and tension over the sheet arc length. (see Section 4.2.11.)

The thickness is given by

$$t = t_0 \exp(-\varepsilon_1) \tag{4.17}$$

The approximate current blank width b in Figure 4.3 can be found by equating volumes. It is sufficiently accurate to determine the thickness at the end of each zone, e.g. at O and A from Equation 4.17, and to calculate the volume in this segment as

$$Vol._{\mathrm{OA}} = R_{\mathrm{P}}\theta_{\mathrm{A}}\frac{t_{\mathrm{O}} + t_{\mathrm{A}}}{2} \tag{4.18}$$

Summing all such volumes from O to F and subtracting from the initial volume gives the current volume between F and G and hence, as there is no change in thickness beyond F_1

$$FGt_0 = b_0 t_0 - \sum_{\mathrm{O}\to\mathrm{F}} \mathrm{Vol.} \tag{4.19}$$

4.2.10 Accuracy of the simple model

A two-dimensional model of a process that is in fact three-dimensional will obviously be an approximation. The magnitude of the errors will depend on the actual process and judgment must be exercised. An additional source of error is the effect of the sheet bending and unbending as it passes around the tool radii, particularly at the die corner radius, at C, in Figure 4.3. Two effects are important. Bending strains will cause work-hardening in the sheet and also, as shown later, bending and unbending under tension reduce the sheet thickness. Both these effects will reduce the strain at which the tension reaches a maximum and can lead to early failure in the side-wall.

4.2.11 (Worked example) 2D stamping

Drawing quality steel of 0.8 mm thickness is formed in a stamping dieas shown in Figure 4.3 but with vertical side walls. The plane strain stress–strain relation is

$$\sigma_1 = 750\varepsilon_1^{0.23}\ \mathrm{MPa}$$

In the two-dimensional plane strain model, the variables are:

Semi punch width:	$a = 330\,\mathrm{mm}$
Punch face radius;	$R_f = 2800\,\mathrm{mm}$
Corner radius:	$R_p,\ R_d = 10\,\mathrm{mm}$
Side wall length	$BC = 28\,\mathrm{mm}$
Land width:	$DE = 0\,\mathrm{mm}$
Clamp width:	$EF = 80\,\mathrm{mm}$

(a) Estimate the blankholder force per side, per unit width to achieve a strain $(\varepsilon_1)_0 = 0.03$ at the centre if the friction coefficient is 0.1.

(b) If s is the arc length measured along the deformed sheet in the above condition, prepare diagrams in which the horizontal axes are each s, and the vertical axes are:
   (i) the membrane strain, ε_1,
   (ii) sheet thickness
   (iii) the tension, T in kN/m, and
   (iv) the contact pressure.

(c) If in the condition shown, the edge of the sheet just comes to the point F, estimate approximately the initial semi blank width.
(d) If, in the position shown, the side wall is about to split, estimate the punch force P and the strain at the centre $(\varepsilon_1)_0$.

Solutions

1. Tension at the centreline

$T_O = \sigma_1 t = K\varepsilon_{1O}^n t_0 \exp(-\varepsilon_{1O})$

$= 750 \times 10^6 \times 0.8 \times (0.03)^{0.23} \exp(-0.03) = 260(\text{kN/m})$

2. Arc length

$$\sin\theta = \frac{a - R_P}{R_f - R_P} = \frac{0.33 - 0.01}{2.8 - 0.01}, \quad \theta = 0.115(\text{rad}) = 6.6^\circ$$

$\text{arc OA} = (R_f - R_p)\theta = (2.8 - 0.01) \times 0.115 = 0.321(\text{m})$

$$\text{arc AB} = R_P\frac{\pi}{2} - \theta = 0.01 \times \frac{\pi}{2} - \theta = 0.015(\text{m})$$

$\text{CB} = 0.028$

$\text{Arc CD} = \pi/2 \times 0.01 = 0.015(\text{m})$

$\text{DF} = 0.080(\text{m})$

$\therefore$ arc length $\text{OF} = 0.460(\text{m})$

3. Tensions

Using Equation 4.13(b)

$T_A = T_O \exp(\mu\theta) = 260 \exp(0.1 \times 0.115) = 263(\text{kN/m})$

$T_B = T_O \exp(\mu\pi/2) = 263 \exp(0.1 \times 3.14/2) = 304(\text{kN/m})$

$T_C = T_B$

$T_C = T_D \exp(\mu\pi/2) = T_D \exp\ (0.1 \times 3.14/2) = 304(\text{kN/m})$

$T_D = T_E = 308/\exp(\mu\pi/2) = 260(\text{kN/m})$

4. Blankholder force

From Equation 4.14, we have

$T_D = 2\mu B,$

$$\therefore B = \frac{T_D}{2\mu} = \frac{260}{2 \times 0.1} = 1300(\text{kN})$$

5. Strains

$T_A = K\varepsilon_{1A}^n t_0 \exp(-\varepsilon_{1A}) = 600\varepsilon_{1A} \exp(-\varepsilon_{1A}) = 263$

$\therefore$ by trial and error, $\varepsilon_{1A} = 0.032$

similarly, $\varepsilon_{1B} = \varepsilon_{1C} = 0.078$

$$\varepsilon_{1D} = \varepsilon_{1E} = 0.032$$

6. Average strains

OA: 0.031; AB: 0.055; BC: 0.078; CD: 0.055; EF: 0.016

7. Original arc length

OA: $L_0 = 0.321\exp(-0.031) = 0.311(\text{m})$

AB: $L_0 = 0.0157\exp(-0.055) = 0.015(\text{m})$

BC: $L_0 = 0.028\exp(-0.078) = 0.026(\text{m})$

CD: $L_0 = 0.0157\exp(-0.055) = 0.015(\text{m})$

DE: $L_0 = 0.080\exp(-0.016) = 0.079(\text{m})$

8. Original blank semi-width

$b = \sum L_0 = 0.446(\text{m})$

Overall average strain $= \ln(0.460/0.446) = 0.031$

9. Ultimate side-wall tension

$T_{ult} = K t_0 n^n \exp(-n) = 340(\text{kN/m})$

10. Maximum centre-line strain

$T_{0\,max} = T_{ult}/\exp(\mu\pi/2) = 290(\text{kN/m})$

$\varepsilon_{10}^{0.23}\exp(-\varepsilon_{10}) = 290/K t_0 = 0.48$

$(\varepsilon_{10})_{max} = 0.05$

11. Pressure

$p = T/R$

$$p_0 = \frac{260 \times 10^3}{2.8} = 0.093(\text{MPa})$$

$$p_A = P_D = \frac{263 \times 10^3}{0.01} = 26.3(\text{MPa}) = P_D$$

$$p_B = P_C = \frac{308 \times 10^3}{0.01} = 31(\text{MPa})$$

$$p_{EF} = B/\text{area} = \frac{1300 \times 10^3}{0.08 \times 1} = 16.3(\text{MPa})$$

12. Thickness

$t = t_0 \exp(-\varepsilon_1)$

$t_A = 0.8 \exp(-0.03) = 0.78(\text{mm})$

$t_B = t_C = 0.74(\text{mm})$

$t_o = t_D = t_E = 0.78(\text{mm})$

4.3 Stretch and draw ratios in a stamping

In stampings of simple shape, it is often useful to determine the *stretch* and *draw ratios* at a section. Figure 4.10 shows the blank and die at the start of the operation. The points D are the tangent points of the sheet at the die corner radius. As the blank is drawn in, the die will mark the sheet slightly so that the position of the material point that was originally at D can be seen on the sheet. The line on the stamping indicated by this marking or scratching is called the *die impact line* and in Figure 4.10(b) it is indicated by the point D′. The length measured around the sheet, $2d$, as shown in Figure 4.10(b), is the current length of $2d_0$ and the *stretching ratio* is defined as

$$SR = \frac{d - d_0}{d_0} \times 100\% \tag{4.20}$$

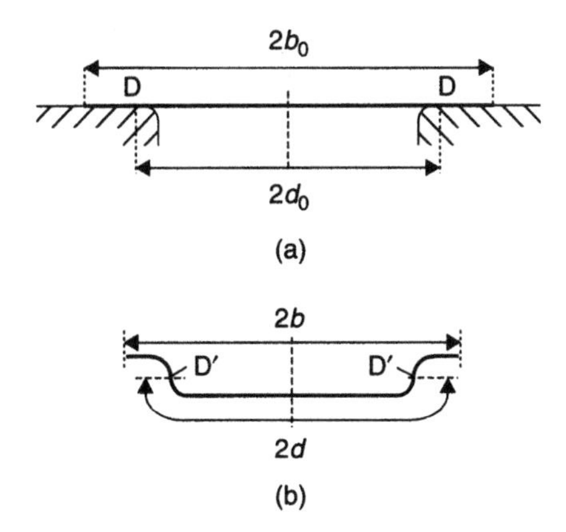

Figure 4.10 Section of a stamping illustrating the drawing and stretching ratios.

The *drawing ratio* is

$$DR = \frac{b_0 - b}{b_o} \times 100\% \tag{4.21}$$

It is often found that problems will occur in stamping if these ratios change too rapidly with successive sections along the tool.

4.4 Three-dimensional stamping model

A number of stampings resemble the rectangular pan shown in Figure 4.11. In the corners, shown shaded, the material is drawn inwards in converging flow. This mode of deformation is considered in a subsequent chapter. The straight sides can be modelled approximately as two-dimensional sections, except that the deformation over the face of the punch is no longer plane strain, but biaxial stretching with a strain ratio in the range $0 < \beta < 1$. Useful information can be determined from a simple model and this is demonstrated in the following worked example.

4.4.1 (Worked example) Stamping a rectangular panel

A rectangular pan as shown in Figure 4.11 is drawn in a die. The base of the panel is flat and it is specified that the base should be stretched uniformly so that a lip of 4 mm height should be drawn up around the edge as shown in Figure 4.12. Determine the side-wall tensions T_1 and T_2 and the punch force required for the following conditions:

initial sheet thickness	0.90 mm
stress strain law	$\bar{\sigma} = 700\,(0.009 + \bar{\varepsilon})^{0.22}$ MPa

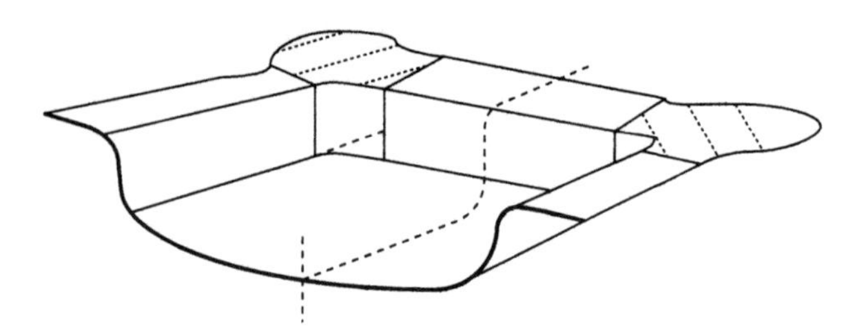

Figure 4.11 One half of a typical rectangular stamping.

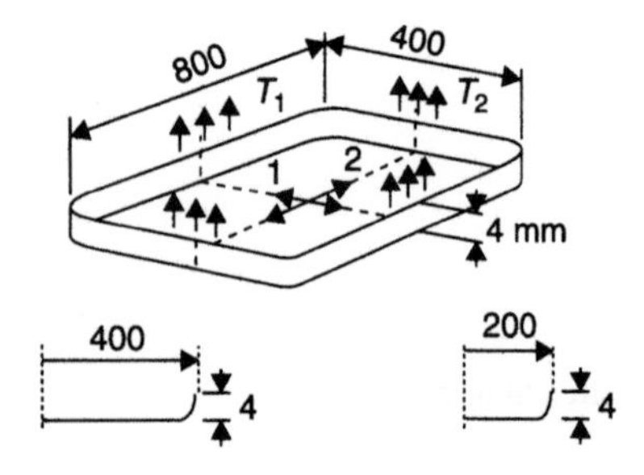

Figure 4.12 Diagram of a rectangular pan drawn to a specified strain.

Solution. For the semi-axes in Figure 4.12(b), the sheet is stretched so that

$$\varepsilon_1 \approx \ln\frac{204}{200} = 0.0198 \quad \text{and} \quad \varepsilon_2 \approx \ln\frac{404}{400} = 0.010 \tag{4.22}$$

the strain ratio is

$$\beta = \frac{\varepsilon_2}{\varepsilon_1} = \frac{0.010}{0.0198} = 0.505 \tag{4.23}$$

and the stress ratio is, from Equation 2.14,

$$\alpha = \frac{2\beta + 1}{2 + \beta} = 0.802 \tag{4.24}$$

From Equations 2.18(b) and 2.19, we obtain

$$\bar{\sigma} = 0.917\sigma_1 \qquad \text{and} \qquad \bar{\varepsilon} = 1.532\varepsilon_1 \tag{4.25}$$

The effective stress at the end of the process is

$$\bar{\sigma} = 700\,(0.009 + 1.532 \times 0.0198)^{0.22} = 344\,\text{MPa}$$

and hence $\sigma_1 = 344/0.917 = 375\,\text{MPa}$ and $\sigma_2 = \alpha.\sigma_1 = 0.802 \times 375 = 301\,\text{MPa}$.

The thickness is

$$t = t_0 \exp\left[-\,(1 + \beta)\,\varepsilon_1\right] = 0.9 \exp\left[-\,(1 + 0.505)\,0.0198\right] = 0.874\,\text{mm}$$

and as $T = \sigma t$,

$$T_1 = 375 \times 0.874 = 326\,\text{kN/m} \quad \text{and} \quad T_2 = 301 \times 0.874 = 263\,\text{kN/m}$$

Summing the tension around the side-wall and neglecting frictional effects, the punch force is, approximately,

$$F = 2\,(326 \times 0.8 + 263 \times 0.4) = 732\,\text{kN} = 0.73\,\text{MN}$$

4.5 Exercises

Ex. 4.1 A material is deforming in plane strain under a major tension of 340 kN/m. The initial thickness is 0.8 mm and the material obeys an effective stress–strain relation $\bar{\sigma} = 700(\bar{\varepsilon})^{0.22}$ MPa. What is the major strain at this point?
[Ans: 0.062]

Ex. 4.2 In the two-dimensional stamping operation shown in Figure 4.13, the side-walls are vertical and the face of the punch is flat. If the blank-holder force B is increased, determine the maximum strain that can be achieved at the centre-line if the coefficient of friction is 0.15 and the sheet obeys the stress strain law $\bar{\sigma} = 600\bar{\varepsilon}^{0.2}$ MPa. What is the blank-holder force required to reach this if the initial sheet thickness is 0.8 mm?
[Ans: 0.026; $2\text{B} = 1780\,(kN/m)$]

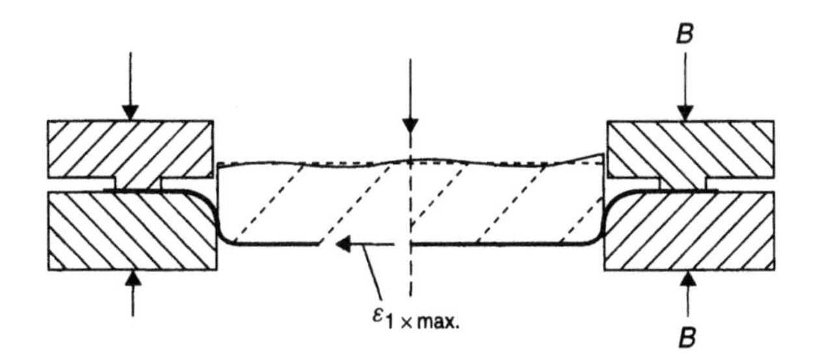

Figure 4.13 Section of a draw die with vertical sidewalls.

Ex. 4.3 For the operation in Ex. 4.2, determine the punch force at the maximum side-wall tension and obtain the ratio of blank-holder force to punch force.
Ans: 676 kN/m; 2.6]

Ex. 4.4 At a point in a stamping process the sheet that is in contact with the punch is shown in Figure 4.3, but the sheet makes an angle of 60° at the tangent point. The strain at the mid-point O is 0.025. The punch has a face radius of 2 m, semi-width of 600 mm, and corner radius of 10 mm. The material obeys an effective stress strain relation of $\overline{\sigma} = 400(\overline{\varepsilon})^{0.17}$ MPa and the initial thickness is 0.8 mm. Determine the tension at O, A and B if the coefficient of friction is 0.10.
[Ans: 197, 203 and 219 kN/m]

5 Load instability and tearing

5.1 Introduction

The previous chapters showed how the plastic deformation of an element can be followed during a metal forming operation. At some instant, the process may be limited or terminated by any one of a number of events and part of the analysis of these operations includes the prediction of process limits. Each process will have its own limiting events and while it may not be easy at first sight to anticipate which event will govern, it is likely to be one of the following.

- *Inability of the sheet to transmit the required force.* In the deep drawing process shown in the Introduction in Figure I.9, the force required to draw the flange inward may exceed the strength of the cup wall. This occurs when the tension (force per unit length) around the circumference reaches a maximum and this will also be seen as a maximum in the punch force. This type of limit is often termed a *global instability* as the whole process must be considered. A similar situation happens in the tensile test. The object of the test is to obtain uniform deformation in the gauge length from which mechanical properties can be measured. When the load reaches a maximum, deformation becomes concentrated in a diffuse neck and is no longer uniform. Actual failure of the strip happens at a later stage, but the uniform deformation process is limited by the load maximum and this is also a global instability.
- *Localized necking or tearing.* The appearance of any local neck that rapidly leads to tearing and failure will obviously terminate a forming operation. This can be considered as *local instability* that can be analysed by considering a local element without involving the whole process.
- *Fracture.* It is possible for a plastically deforming element to fracture in almost a brittle manner. This is not common in sheet used for forming and is often preceded by some local instability. Some instances where fracture may have to be considered will be mentioned in this chapter.
- *Wrinkling.* If one principal stress in an element is compressive, the sheet may buckle or wrinkle. This is a *compressive instability* and resembles the buckling of a column. In sheet it is difficult to predict and will not be studied here.

In a complicated process such as stamping an irregular part, the overall punch force is made up of the loads created in forming a number of different regions. It may not be possible to identify the onset of a load maximum in any one region and even if the total load reaches a maximum, this may not constitute a limit in the process. Most forming machines are *displacement controlled* and the motion of the punch is not determined by the load acting on it, but by the mechanism that drives it. In other words, the machine will not go out of control because the punch force versus displacement characteristic reaches a maximum.

Global instabilities such as in deep drawing are predicted by analysing the process as a whole and will be examined in sections dealing with these operations. On the other hand, local instabilities, as the name implies, can be understood by studying the deformation of a single element. In sheet forming, the instability that is most likely to occur is the sudden growth of a local neck leading to tearing. Under some conditions local necking will occur when the tension reaches a maximum, but in other conditions, a tension maximum may not be associated with the sudden growth of a local neck or catastrophic failure.

5.2 Uniaxial tension of a perfect strip

We first consider the theoretical case of a parallel strip of metal, as in the gauge length of a tensile test-piece. We consider that the properties are uniform throughout and the geometry is perfect. When this is stretched in tension as shown in Figure 5.1, the volume remains constant and the following relations apply. The cross-sectional area is $A = wt$ and the volume is

$$Al = A_0 l_0 \tag{5.1}$$

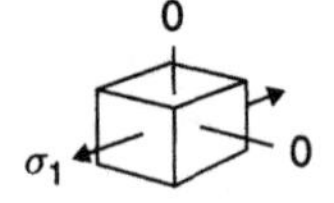

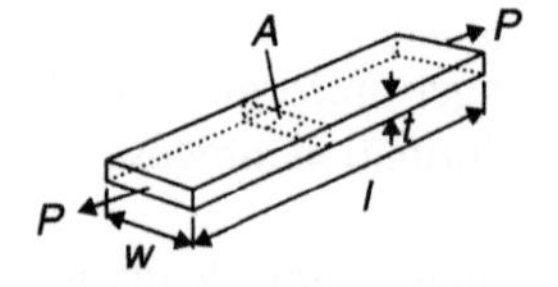

Figure 5.1 Diagram of a perfect strip deformed in uniaxial tension.

Differentiating Equation 5.1, we obtain

$$\frac{dA}{A} + \frac{dl}{l} = 0 \qquad \text{or} \qquad \frac{dl}{l} = d\varepsilon_1 = -\frac{dA}{A} \tag{5.2}$$

The strain in the strip is

$$\varepsilon_1 = \ln\frac{l}{l_0} \tag{5.3}$$

and the stress is

$$\sigma_1 = \frac{P}{A} = \frac{P}{A_0}\frac{l}{l_0} \tag{5.4}$$

The load in the strip is $P = \sigma_1 A$; as the strip deforms, σ_1 will increase for a strain-hardening material and the cross-sectional area will decrease, i.e. $d\sigma_1$ will always be positive and dA will be negative. At some stage, the rate of strain-hardening will fall below the rate of reduction in area and the load will reach a maximum. At this instant,

$$dP = d(\sigma_1 A) = 0$$

or

$$\frac{dP}{P} = \frac{d\sigma_1}{\sigma_1} + \frac{dA}{A} = 0 \tag{5.5}$$

Combining with Equation 5.2, we obtain the condition for maximum load as

$$\frac{1}{\sigma_1}\frac{d\sigma_1}{d\varepsilon_1} = 1 \tag{5.6}$$

The function on the left-hand side of Equation 5.6 is a material property that is known as the *non-dimensional strain-hardening* characteristic and it could be determined from a material test. If the material obeys a simple power law, $\sigma_1 = K\varepsilon_1^n$, similar to that introduced in Section 3.5.1, this function is

$$\frac{1}{\sigma_1}\frac{d\sigma_1}{d\varepsilon_1} = \frac{nK}{\sigma_1}(\varepsilon_1)^{n-1} = \frac{n}{\varepsilon_1} \tag{5.7}$$

The form of this curve is illustrated in Figure 5.2.

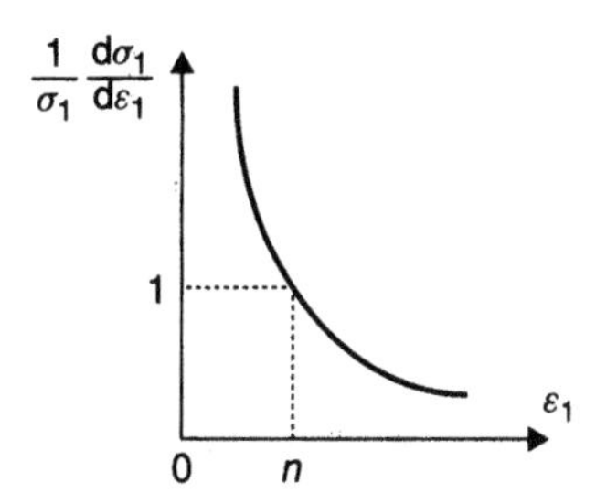

Figure 5.2 Non-dimensional strain-hardening curve for a power law material.

Combining Equations 5.6 and 5.7,

$$\frac{1}{\sigma_1}\frac{d\sigma_1}{d\varepsilon_1} = \frac{n}{\varepsilon_1} = 1 \tag{5.8}$$

and the strain at maximum load is

$$\varepsilon_1^* = n \tag{5.9}$$

where the star denotes the maximum load condition. This point in the diagram is illustrated in Figure 5.2.

Equation 5.6 is the well-known Considere condition for maximum load in a bar or strip in tension. If the bar is part of a load-carrying structure, then this is a significant event. It is often stated as the condition for the start of diffuse necking in a tensile strip. This may be the case in a real strip, but in a perfect strip, unlike an actual test-piece, diffuse necking in one region cannot occur as all elements in a perfect strip must behave in an identical fashion and so uniform deformation would always exist.

For a perfect strip in which the material obeys the power law expression for the stress–strain curve, the load to deform the strip may be calculated, i.e.,

$$P = \sigma_1 A = K\varepsilon_1^n A_0 \frac{l_0}{l} = K A_0 \varepsilon_1^n \exp(-\varepsilon_1) \tag{5.10}$$

The load in Equation 5.10 for a given material is a function of material properties, initial cross-sectional area and axial strain and is plotted as a function of strain in Figure 5.3.

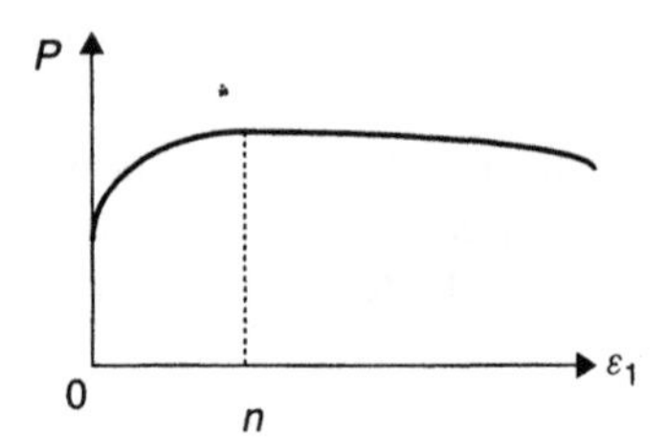

Figure 5.3 Variation of load with strain in a perfect strip.

If we compare the theoretical load–strain curve for a perfect strip as shown in Figure 5.3 with that from a real tensile strip, as in Figure 1.2, we see there is a significant difference. In a real specimen, the strip starts to neck at the maximum load and the load then falls off much more rapidly than in Figure 5.3. In order to understand the actual phenomenon in a real strip, we must include an imperfection in the strip and then analyse the deformation.

5.3 Tension of an imperfect strip

We consider a tensile strip in which a slight imperfection exists. This can be characterized by a short region having initially a slightly smaller cross-sectional area; i.e. if the initial area of most of the strip is A_0, then the imperfection is initially of area $(A_o + dA_0)$ where dA_0 is a small negative quantity. At some stage in the deformation, the strip will be as illustrated in Figure 5.4.

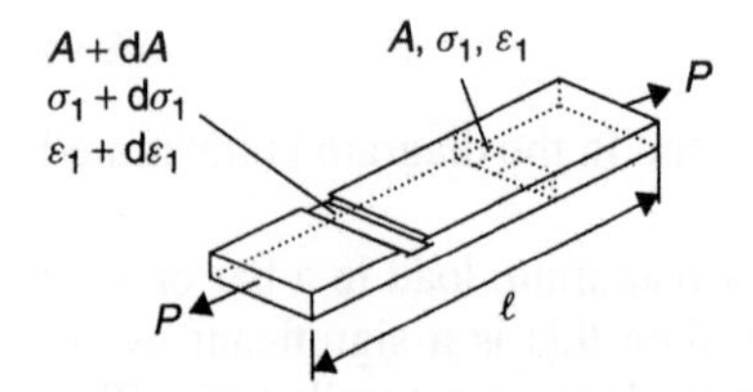

Figure 5.4 Tension of an imperfect strip.

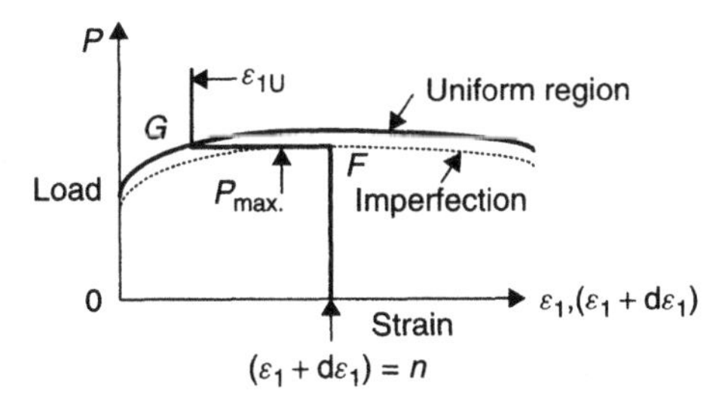

Figure 5.5 Load, strain diagram for the uniform region and the imperfection in a tensile strip.

The same load is transmitted by the uniform region and the imperfection and, following Equation 5.10, this can be written as

$$P = KA_0\varepsilon_1^n \exp(-\varepsilon_1) = K(A_0 + dA_0)(\varepsilon_1 + d\varepsilon_1)^n \exp[-(\varepsilon_1 + d\varepsilon_1)] \qquad (5.11)$$

The load versus strain curves for each region are illustrated in Figure 5.5, noting that the strain in the imperfection is $(\varepsilon_1 + d\varepsilon_1)$.

The imperfection will reach a maximum load, following Equation 5.9, when the strain is $(\varepsilon_1 + d\varepsilon_1) = n$. At this load, the uniform region is only strained to the point G and the strain is $\varepsilon_{1U} < n$; this is known as the maximum uniform strain as it is the strain measured in the uniform region after the test. It may be seen, from Figure 5.5 that the uniform region cannot strain beyond the point G as it would require a higher load than can be transmitted by the imperfection. If the test is continued beyond the maximum load $P_{max.}$, only the imperfection will deform and it will do this under a falling load. All real tension strips will contain some imperfections even if they are very slight. The greatest imperfection will become the site of the diffuse neck and once the maximum load-carrying capacity is reached in the imperfection or neck, all the deformation is concentrated in the neck and the uniform region will unload elastically as the load falls. The diffuse neck that is seen in actual test-pieces usually extends for a distance approximately equal to the width of the strip and its deformation will only contribute a small amount to the overall elongation. The curve of load versus total elongation, therefore, will fall more rapidly than for a perfect strip, as shown in Figure 5.6. (It should be noted that the post-uniform elongation will also be dependent on the ratio of the gauge length to the strip width. Other things being equal, if the gauge length is much larger than the strip width, the length of the diffuse neck will be a small fraction of the gauge length and the post-uniform extension will be small. For tests in which the gauge length is more nearly equal to the strip width, the elongation will be greater.)

The difference between the maximum strain in the uniform region of an imperfect strip and the strain $\varepsilon_1 = n$ at maximum load in a perfect strip can be found approximately for a material obeying the power law model $\sigma_1 = K\varepsilon_1^n$. If, in Equation 5.11, we substitute $\varepsilon_1 + d\varepsilon_1 = n$ for the strain at maximum load in the imperfection and ε_U for the strain in the uniform region, then we obtain

$$\left(\frac{\varepsilon_U}{n}\right)^n \exp(n - \varepsilon_U) = 1 + \frac{dA_0}{A_0} \qquad (5.12)$$

In real cases, both $(n - \varepsilon_U)$ and dA_o/A_0 will be small quantities. It is shown in a Note at the end of this chapter that by taking the first terms only in the series expansions for the

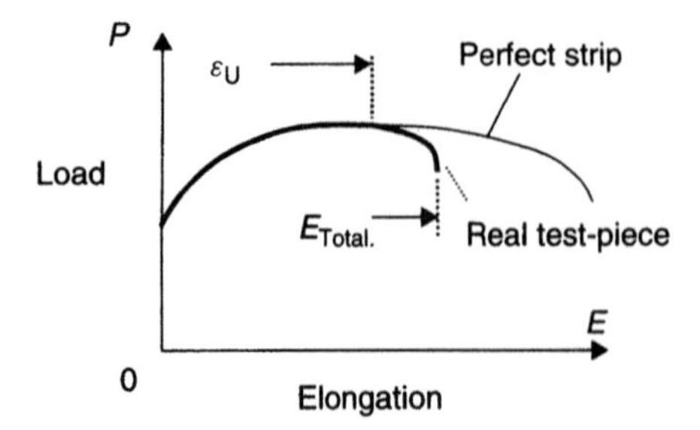

Figure 5.6 Load extension diagram for a perfect strip and one containing an imperfection.

functions in Equation 5.12, we obtain

$$(n - \varepsilon_U) \approx \sqrt{-n\frac{dA_o}{A_0}} \tag{5.13}$$

As dA_0 is a negative quantity, the root of Equation 5.13 is real and the difference between the maximum strain in the uniform region and the strain-hardening index n is dependent on the strain-hardening index and the magnitude of the imperfection.

5.3.1 (Worked example) maximum uniform strain

The length, width and thickness of the parallel reduced section of a tensile test-piece are 100, 12.5 and 0.8 mm respectively. The material has a stress, strain curve fitted by the relation $\sigma_1 = 700\varepsilon_1^{0.22}$ MPa. Over a small length, the width is 0.05 mm less than elsewhere. Estimate the strain in the uniform region of the test-piece after the strip has been tested to failure.

Solution. The initial cross-sectional area is $0.8 \times 12.5 = 10\,\text{mm}^2$. The difference in area at the imperfection is $0.8 \times 0.05 = 0.04\,\text{mm}^2$. The imperfection ratio is $dA_0/A_0 = -0.004$. From Equation 5.13,

$$0.22 - \varepsilon_U = \sqrt{-0.22\,(-0.004)} = 0.03 \qquad \text{or} \qquad \varepsilon_U = 0.19 \tag{5.14}$$

If the imperfection did not exist, the uniform strain would be 0.22. We thus see that an imperfection of only 0.4%, reduces the maximum uniform strain by $(0.03/0.22) \times 100 = 13.5\%$. This demonstrates a phenomenon often observed in sheet metal forming: *very small changes in initial conditions can give large changes in the final result.* It is also found that if repeated tests are performed on apparently uniform material, there is considerable scatter in the observed maximum uniform strain. This is probably due to differences in the magnitude of imperfections that exist in individual test-pieces.

5.3.2 The effect of rate sensitivity

The analysis above assumes that the material strain-hardens, but is insensitive to changes in strain rate. In sheet metal at room temperature, rate sensitivity is small, but it can affect necking as strain rates in the neck can become quite high when uniform straining ceases and deformation becomes concentrated in necks. To illustrate the effect, we consider here a non-strain-hardening material having a tensile stress–strain rate curve

$$\sigma_1 = B\dot{\varepsilon}_1^m \tag{5.15}$$

where $\dot{\varepsilon}_1$ is strain rate, i.e.

$$\dot{\varepsilon}_1 = \frac{d\varepsilon_1}{dt} = \frac{dl/l}{dt} = \frac{v}{l} \tag{5.16}$$

In Equation 5.16, t is time, v the velocity or cross-head speed of the testing machine and l is the length of the parallel reduced section of the test-piece.

If we consider an imperfect strip as illustrated in Figure 5.4, then the force transmitted through both regions is

$$P = \sigma_1 A = (\sigma_1 + d\sigma_1)(A + dA) \tag{5.17}$$

from which we obtain

$$\frac{d\sigma_1}{\sigma_1} = -\frac{dA}{A} \tag{5.18}$$

Thus the difference in stress for the two regions is proportional to the magnitude of the imperfection. Differentiating the material law, Equation 5.15, we obtain

$$\frac{d\dot{\varepsilon}_1}{\dot{\varepsilon}_1} = \frac{1}{m}\frac{d\sigma_1}{\sigma_1} = -\frac{1}{m}\frac{dA}{A} \tag{5.19}$$

This indicates that for a given imperfection, the difference in the strain rate between the imperfection and the uniform region is inversely proportional to *the strain-rate sensitivity index* m. If m is small, as it is in most sheet metal at room temperature, the difference in strain rate will be large and the imperfection will grow rapidly. In a special class of alloys, termed *superplastic*, m is unusually high, about 0.3; strips of these materials can be extended several hundred per cent as the growth of imperfections is very gradual. In some non-metals, such as molten glass, $m \sim 1$, and these can be drawn out almost indefinitely.

In materials with low values of m, rate sensitivity will not greatly influence the maximum uniform strain, because, as shown in Figure 5.5, the strain and hence strain rate are approximately similar in both regions up to maximum load. Beyond this point, the necking process will be affected by rate sensitivity and it is found that the post-uniform elongation is higher in materials with greater rate sensitivity.

5.4 Tensile instability in stretching continuous sheet

It is shown in the previous section that in a tensile strip, provided some imperfections exist, diffuse necking will start when the load reaches a maximum. On the other hand, in sheet that is stretched over a punch, diffuse necking is not observed. The tension in the sheet may reach a maximum, but the punch will exert a geometric constraint on the strain distribution that can develop. This is illustrated in Figure 5.7. If a diffuse neck did develop, the increased strain would lead to the sheet moving away from the punch and this is implausible. It is possible that the strain may accelerate in some region compared with

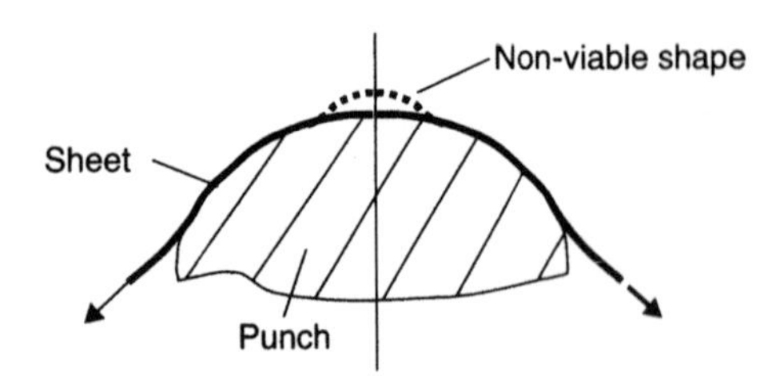

Figure 5.7 Diagram illustrating that a diffuse neck would lead to a non-viable strain distribution in a continuous sheet.

another, but in a displacement controlled machine, this would not constitute an end-point in the process.

Observation shows that in continuous sheet, local necks do develop similar to those that occur within the diffuse neck of a tensile test-piece. The width of these local necks is roughly equal to the thickness of the sheet and they will not influence the overall or global strain distribution. However, they lead very quickly to tearing of the sheet and the end of the process. In designing sheet forming processes therefore, it is important to understand the conditions under which these local necks develop. The theory of local necking is not fully developed, but a model of necking is given here that does predict many of the observed phenomena.

We consider a region of the sheet deforming uniformly in a proportional process as shown in Figure 5.8. The deformation in this region may be specified as

$$\begin{aligned} &\sigma_1; \quad \sigma_2 = \alpha\sigma_1; \quad \sigma_3 = 0 \\ &\varepsilon_1; \quad \varepsilon_2 = \beta\varepsilon_1; \quad \varepsilon_3 = -(1+\beta)\,\varepsilon_1 \end{aligned} \tag{5.20}$$

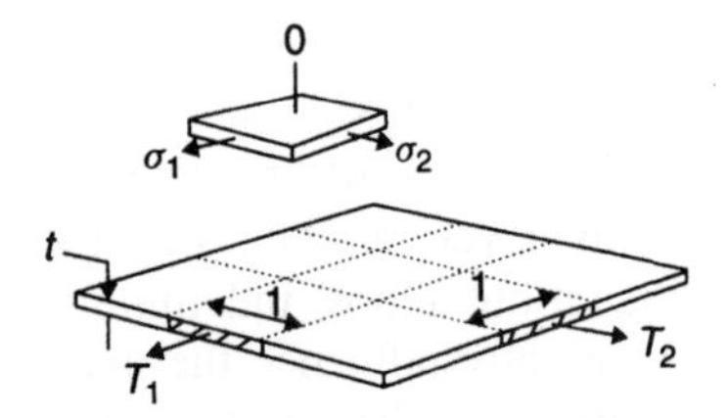

Figure 5.8 Uniform deformation of part of a continuous sheet in a plane stress proportional process.

The principal tensions in the sheet are

$$T_1 = \sigma_1 t \quad \text{and} \quad T_2 = \alpha T_1 = \sigma_2 t \tag{5.21}$$

These tensions will remain proportional during forming.

5.4.1 A condition for local necking

The condition postulated for local necking is that it will start when the major tension reaches a maximum. As the process is proportional, α and β will be constant. Differentiating Equation 5.21, we obtain

$$\frac{dT_1}{T_1} = \frac{d\sigma_1}{\sigma_1} + \frac{dt}{t} = \frac{d\sigma_1}{\sigma_1} + d\varepsilon_3 = \frac{d\sigma_1}{\sigma_1} - (1+\beta)\,d\varepsilon_1 \tag{5.22}$$

When the tensions reach a maximum, Equation 5.22 becomes zero and the non-dimensional strain-hardening is,

$$\frac{1}{\sigma_1}\frac{d\sigma_1}{d\varepsilon_1} = 1 + \beta \tag{5.23}$$

This is only valid for $\beta > -1$. If $\beta < -1$, the sheet will thicken and for a strain-hardening material, the tension will never reach a maximum.

For a material that obeys the generalized stress–strain law

$$\overline{\sigma} = \sigma_f = K(\overline{\varepsilon})^n \tag{5.24}$$

it may be shown by substituting Equations 2.18(b) and 2.19(c) in Equation 5.24 that the relationship between principal stress and strain during proportional deformation in which α and β are constant may be written as

$$\sigma_1 = K'\varepsilon_1^n \tag{5.25}$$

where K' is a material constant that can be calculated from K, n, α and β.

Differentiating Equation 5.25 shows that

$$\frac{1}{\sigma_1}\frac{d\sigma_1}{d\varepsilon_1} = \frac{n}{\varepsilon_1}$$

and substituting in Equation 5.23, we obtain

$$\varepsilon_1^* = \frac{n}{1+\beta} \qquad \text{and} \qquad \varepsilon_2^* = \frac{\beta n}{1+\beta}$$

or

$$\varepsilon_1^* + \varepsilon_2^* = n \tag{5.26}$$

where the star indicates the strain at maximum tension. From Equation 5.26, the line in the strain diagram representing maximum tension is plotted in Figure 5.9(a); this has a slope of 45 degrees and intercepts the major strain axis at a value of n.

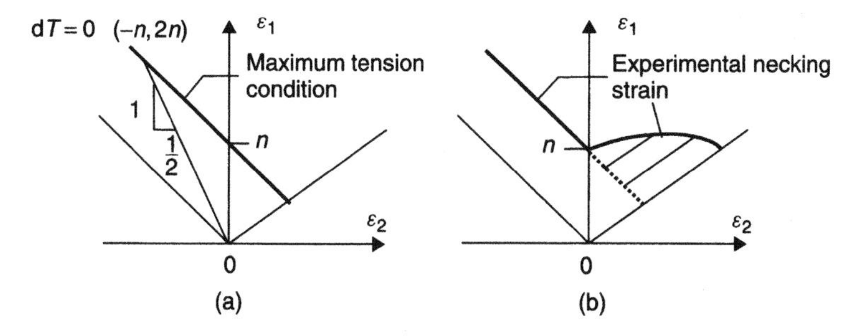

Figure 5.9 (a) Strains at the maximum tension in a continuous sheet. (b) Experimentally observed necking strains in sheet.

In the tensile test, $\beta = -1/2$, and maximum tension occurs when

$$\frac{1}{\sigma_1}\frac{d\sigma_1}{d\varepsilon_1} = \frac{1}{2} = \frac{n}{\varepsilon_1} \qquad \text{or} \qquad \varepsilon_1^* = 2n$$

If the maximum tension condition does signify the onset of local necking as hypothesized, then the local necking strain in uniaxial tension is $\varepsilon_1^* = 2n$; this is twice the strain for the load maximum and the start of diffuse necking as given by Equation 5.9. This agrees with observation for low carbon steel sheet. As the diffuse neck develops in a tensile strip, the deformation in the neck will be approximately uniaxial tension and after some further straining a local neck will develop in the diffusely deforming neck.

We now compare the strain at maximum tension with experimentally determined strains at the onset of local necking. These strains could be measured from grid circles close to a local neck in a formed part. If strains are measured for many different strain paths in both quadrants of a strain diagram, such as Figure 5.9(a), it is possible to establish a 'Forming Limit Curve' (FLC) that delineates the boundary of uniform straining and the onset of local necking. (In practice, there is some scatter in measured necking strains and instead of a single curve there is a band within which necking is likely to occur. Here we will consider the limit as a single curve.) For materials with a similar strain-hardening index n it is found that in the second quadrant of Figure 5.9(a), i.e. for strain paths in which $\beta < 0$, the experimental Forming Limit Curve is approximately coincident with the maximum tension line. In some cases it lies a little above the line shown; the difference may be due to the effect of sheet thickness, friction, contact pressure and grid circle size on the necking process and its measurement. As a first approximation, we conclude that the maximum tension criterion does provide a reasonable theoretical model for local necking strains in the second quadrant ($\beta < 0$).

If both the major and minor strains are positive and the process is one of biaxial stretching with $\beta > 0$, the experimental Forming Limit Curve does not follow the maximum tension line. A typical example is shown in Figure 5.9(b). This suggests that in this quadrant there is some process that stabilizes or slows down necking after the tension has reached a maximum, and this is examined in a later section.

We anticipate that the local neck in Figure 5.10 would occur along a line of pre-existing weakness at a limiting strain in the uniform region that is approximately that given in Equation 5.26. If we identify the uniform region as A and the imperfection as B then certain conditions have been assumed in the analysis of the necking process. These are:

- the stress and strain ratios must remain constant, as assumed in the differentiation, both before and during the necking process;
- for the process to be a local one, the necking process should not affect the boundary conditions in Figure 5.10.

The second condition ensures that the neck must take the form of a narrow trough in the sheet, as in Figure 5.10, rather than as a patch or diffuse region that would influence conditions away from the neck. Once the necking process becomes catastrophic, in the sense that the uniform region A ceases to strain, the strain increment parallel to the neck, in the y direction in Figure 5.10, will be zero. Geometric constraint requires that the strain increment along the neck must be equal to that in the same direction just outside it; i.e. the strain increment in the y direction in both regions A and B along the neck, must be zero. The first condition above requires that the strain ratio does not change, the second that it

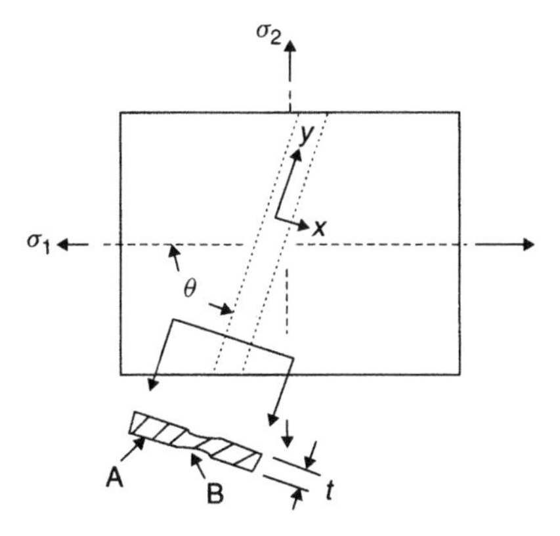

Figure 5.10 A local neck formed in a continuous sheet oriented at an angle θ to the maximum principal stress.

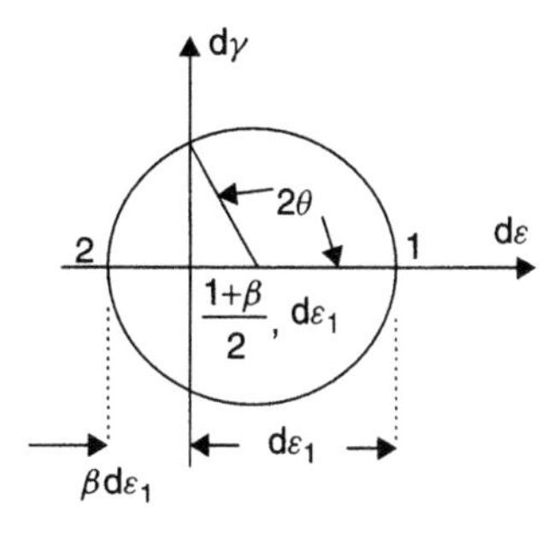

Figure 5.11 Mohr circle of strain increment to determine the angle of zero extension.

is zero during necking, therefore the strain increment in the, y,direction must be zero at all times, i.e. *the neck can develop only along a direction of zero extension.* This direction of the neck can be found from the Mohr circle of strain increment shown in Figure 5.11.

The centre of the circle is at

$$\frac{1+\beta}{2}\,\mathrm{d}\varepsilon_1$$

and the radius of the circle is

$$\frac{1-\beta}{2}\,\mathrm{d}\varepsilon_1$$

The direction of zero extension, $\mathrm{d}\varepsilon_y = 0$, is given by

$$\cos 2\theta = \frac{1+\beta}{1-\beta} \tag{5.27}$$

For uniaxial tension, $\beta = -1/2$, we find that the angle the neck makes is $\theta = 55^\circ$ and for plane strain, $\beta = 0$, the neck is perpendicular to the maximum principal stress, $\theta = 90^\circ$. If $\beta > 0$, there is no direction in which the extension is zero.

The analysis above shows that if there is a direction in which there is no extension, local necking along the direction of zero extension is possible when the tension reaches a maximum and the differentiation as in Equation 5.22 where α and β are constant is valid. If there is no direction of zero extension, for example in a stretching process in which $\beta > 0$, the strains at which the tension is a maximum are still given by Equation 5.26, but geometric constraints prevent the instantaneous growth of local necks. Therefore in Figure 5.9(a), the line shown predicts maximum tension in both quadrants, but only indicates the onset of local necking in the second quadrant in which the minor strain is negative. The diagrams, Figure 5.9 (a) and (b) also suggest that at plane strain where $\beta = 0$ and $\varepsilon_2 = 0$, the major strain at necking is a minimum.

5.4.2 Necking in biaxial tension

In the first quadrant of the strain diagram where both principal strains are positive or tensile, there is no direction of zero extension and, as discussed above, necking of the type illustrated in Figure 5.10 is not possible. Experimentally it is observed that necking still occurs under biaxial tension, but as shown in Figure 5.9(b), at a strain greater than the attainment of maximum tension and usually along a line perpendicular to the major tensile stress. To explain this, a different model is required and this is outlined below. It is necessary to assume some pre-existing defect in the sheet and, for simplicity, we shall consider a small imperfection perpendicular to the greatest principal stress as illustrated in Figure 5.12.

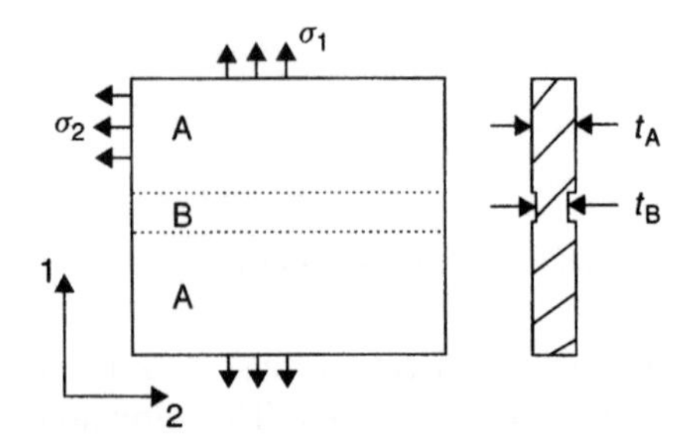

Figure 5.12 An imperfection B in a uniform region A of a sheet deforming in biaxial strain.

The imperfection is a groove B in which the thickness t_B is initially slightly less than that in the uniform region t_A and characterized by an inhomogeneity factor

$$f_o = \left(\frac{t_B}{t_A}\right)_0 \tag{5.28}$$

A typical value of this inhomogeneity would be of the order of $(1 - f_0) = 0.001$. Strain in the region B parallel to the groove would be constrained by the uniform region A so that a compatibility condition is

$$\varepsilon_{2B} = \varepsilon_{2A} \tag{5.29}$$

We investigate a proportional deformation process for the uniform region specified by

$$\begin{aligned} &\sigma_{1A}; \quad \sigma_{2A} = \alpha_0 . \sigma_{1A}; \quad \sigma_{3A} = 0 \\ &\varepsilon_{1A}; \quad \varepsilon_{2A} = \beta_0 . \varepsilon_{1A}; \quad \varepsilon_{3A} = -(1 + \beta_0)\,\varepsilon_{1A} \end{aligned} \tag{5.30}$$

The same tension in the 1 direction is transmitted across both regions, therefore

$$T_1 = \sigma_{1A} t_A = \sigma_{1B} t_B \tag{5.31}$$

and consequently

$$\frac{t_B}{t_A} = \frac{\sigma_{1A}}{\sigma_{1B}} = f \tag{5.32}$$

For given initial conditions and stress strain curve, the deformation of both regions in Figure 5.12 can be analysed numerically. This will not be done here, but the salient features of such an analysis will be illustrated. We consider the initial yielding as shown in Figure 5.13. If the material has a definite initial yield point $(\sigma_f)_0$, then on loading, the groove will reach yield first as, from Equation 5.32, $\sigma_{1B} > \sigma_{1A}$ for $f < 1$. The material in the groove cannot deform because of the geometric constraint, Equation 5.29, therefore as the stress in A increases to reach the yield locus, the point representing the region B must move around the yield locus to B_0 as shown.

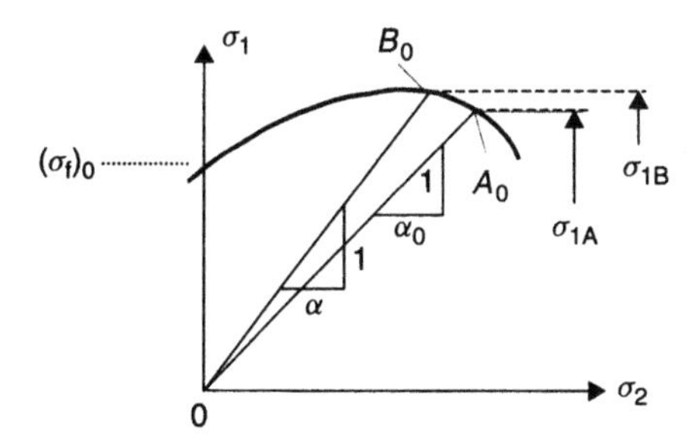

Figure 5.13 Initial yielding conditions for the uniform region and the imperfection.

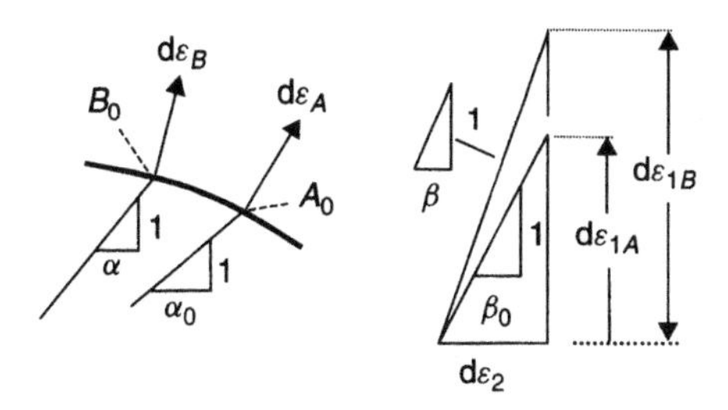

Figure 5.14 Strain vectors for the imperfection and the uniform region.

We now consider some increment in deformation, for which, from Equation 5.29, the increments parallel to the groove must be the same, i.e. $d\varepsilon_{2A} = d\varepsilon_{2B}$. The strain vectors for both regions are illustrated in the magnified view of the yield locus, Figure 5.14, noting that these strain vectors are perpendicular to the yield surface. Because each region is now deforming under different stress and strain ratios, we note that the strain vector for the groove has rotated to the left and for the same strain increment parallel to the groove, the strain increment across the groove will be greater than that in the uniform region A and the inhomogeneity will become greater, i.e. f will diminish. As shown in the insert on the right, $d\varepsilon_{1B} > d\varepsilon_{1A}$.

A numerical analysis will show that the strain in the groove will run ahead of that in the uniform region, but only slightly while the tension is increasing. The effect gradually accelerates after the tension maximum and continues until the groove reaches a state of plane strain as shown in Figure 5.15.

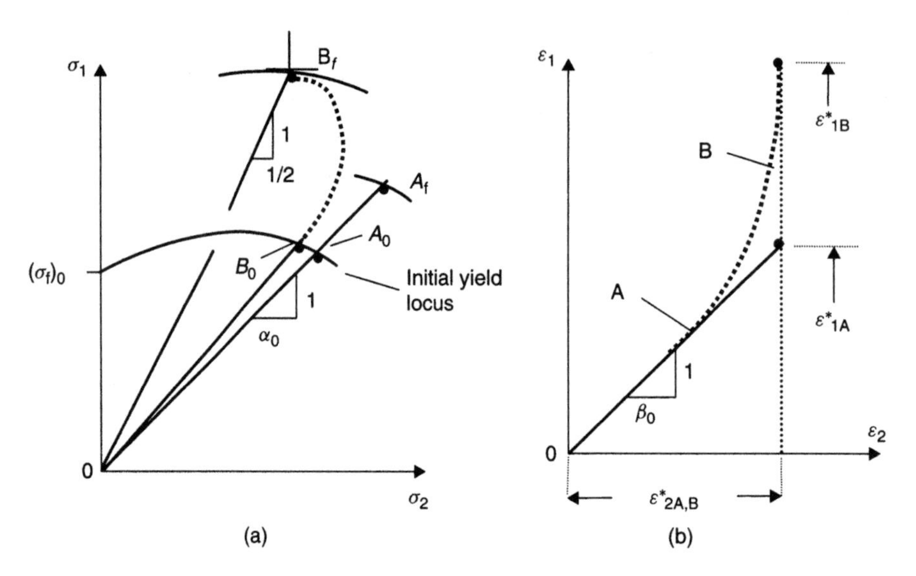

Figure 5.15 Growth of an imperfection from initial yielding to the limit at plane strain in (a) the stress diagram and (b) the strain diagram.

When the stress state in the groove reaches plane strain, at B_f in Figure 5.15(a), the strain parallel to the groove ceases. The groove will then continue until failure (tearing) and the strain in the uniform region ceases. This strain state, just outside the neck, is the maximum strain that can be achieved in this process (for these values of α_0 and β_0) and the strains, ε_{1A}^* and ε_{2A}^* are known as the *limit strains.*

If the analysis is repeated for different values of α_0 and β_0, a diagram can be established in the biaxial strain region as illustrated in Figure 5.16. Combining this curve with the maximum tension line in the second quadrant, we obtain a curve indicating the onset of local necking in both regions. This is known as the *Forming Limit Curve* and is a valuable material property curve; it is used frequently in failure diagnosis of sheet metal forming. The shape of the curve depends on a number of different material properties and on the initial inhomogeneity chosen, i.e. f_o. The homogeneity factor cannot be determined independently, but for small values of $(1 - f_o)$, the curve will intersect the major strain axis at about the value of the strain-hardening index n.

The forming limit curve describes a local process, necking and tearing, that is a material property curve dependent on the strain state, but not on the boundary conditions. The object of sheet metal process design, therefore, is to ensure that strains in the sheet do not approach this limit curve. For the process to be robust and able to tolerate small changes in material or process conditions, a safe forming region can be identified that has a suitable

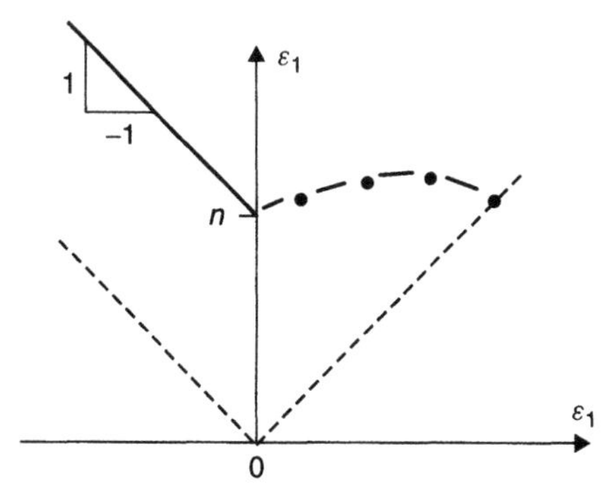

Figure 5.16 The forming limit diagram satisfying the maximum tension criterion on the left-hand side of the diagram and that derived from the imperfection analysis on the right.

margin between it and the limit curve. Different materials will have different forming limit curves and in the following section the effect of individual properties will be outlined.

5.5 Factors affecting the forming limit curve

5.5.1 Strain-hardening

As shown above, the forming limit curve intercepts the major strain axis at approximately the value of the strain-hardening index n. As n decreases, the height of the curve will also decrease as shown in Figure 5.17. Processes in which biaxial stretching is required to make the part usually demand fully annealed, high n sheet; unfortunately, materials with a high n usually have a low initial strength. Many strengthening processes, particularly cold-working, will drastically reduce n and this will make forming more difficult. It is found that as $n \to 0$, the plane strain forming limit along the vertical axis will tend to zero, however, along the equal biaxial direction (the right-hand diagonal) for which $\varepsilon_1 = \varepsilon_2$, the forming limit is not zero and fully cold-worked sheet can be stretched in biaxial tension, but not in any other processes. Except in drawing processes with high negative minor strain, i.e. $\varepsilon_2 \approx -\varepsilon_1$, strain-hardening is usually the most important factor affecting formability.

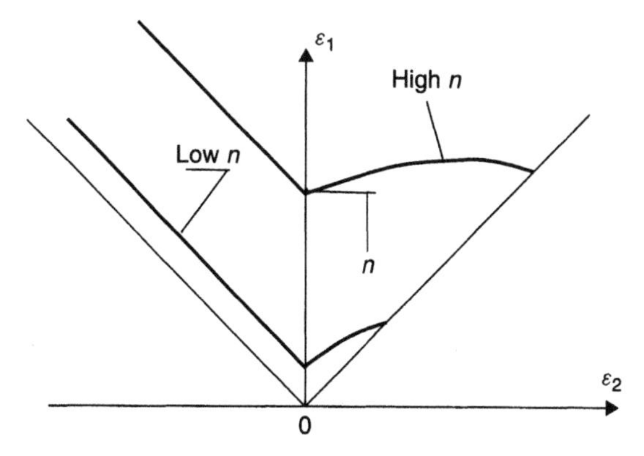

Figure 5.17 Forming limit curves for a high and a low strain-hardening sheet.

5.5.2 Rate sensitivity

It was shown above that in the tensile test, rate sensitivity will not affect the strain at which the tension reaches a maximum, but it will influence the rate of growth of a neck. In biaxial stretching, it has been shown that necking is a gradual process beyond the maximum tension condition and is controlled by the shape of the yield locus. In this region, rate sensitivity will delay growth of the neck as shown in Figure 5.18(a). As shown in Figure 5.18(b), the forming limit curve for a material with a high rate sensitivity could intercept the major strain axis at a strain greater than n.

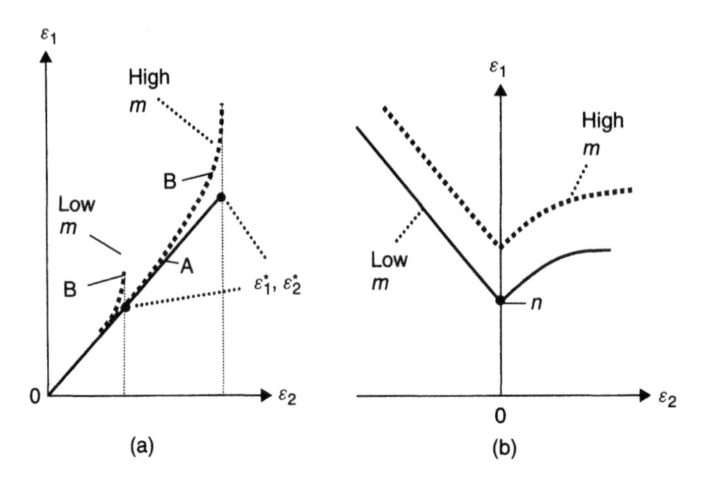

Figure 5.18 Diagram showing the effect of rate sensitivity on (a) the rate of growth of a neck and (b) on the forming limit curve.

5.5.3 Ductile fracture

In the discussion above, it was assumed that tearing in the sheet came about after the necked region B had reached a state of plane strain and that this neck would then proceed to failure without further straining in the uniform region A. In many ductile materials, this is the case and the actual strain at which the neck fractures will not influence the limit strain in A. In less ductile materials, the material within the neck may fracture before plane strain is reached as shown in Figure 5.19; this will reduce the limit strains ε_{1A}^* and ε_{2A}^*. Fracture in ductile materials often results from intense localization of strain on planes of maximum shear. It is possible to measure these fracture strains and plot them on a strain diagram similar to the forming limit. If these fracture curves are well away from the forming limit curve, it may reasonably be assumed that they will not influence the limit strains.

5.5.4 Inhomogeneity

As mentioned, inhomogeneity has not been well characterized in typical sheet. It may be expected that the greater the imperfection, the lower will be the limit strain (Figure 5.20), so that with large imperfections, the plane strain limit strain may be less than the strain-hardening index n. In this work, the imperfection has been expressed in terms of a local

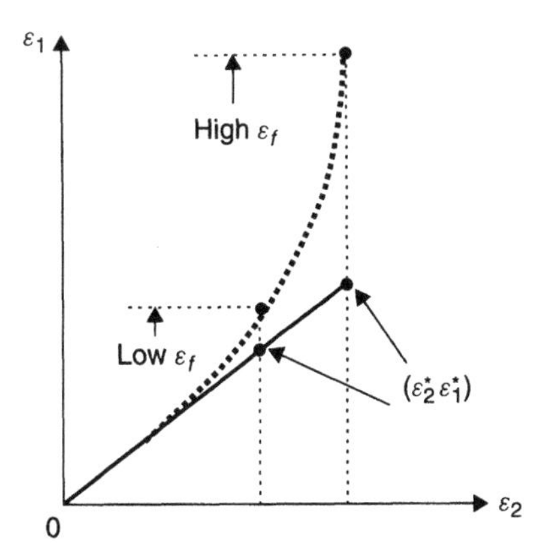

Figure 5.19 Diagram showing the effect of fracture strain ε_f on the limit strains.

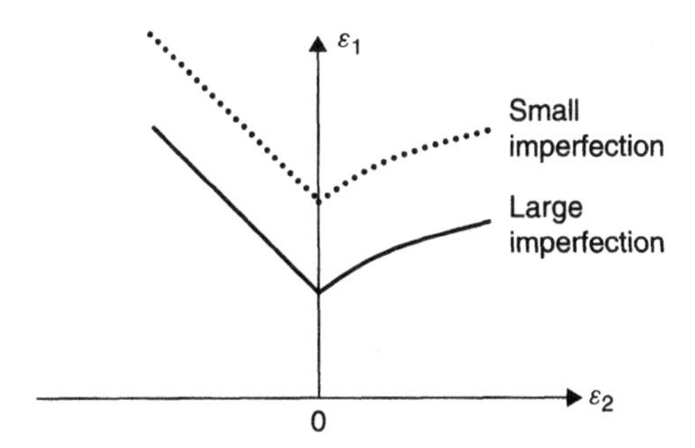

Figure 5.20 Effect of the magnitude of the imperfections on the forming limit curves.

reduction in thickness, but other forms of imperfection are possible, such as inclusions, local reductions in strength due to segregation of strengthening elements or texture variation. Surface roughness may also be a factor. Whatever the form of the imperfection, it will also have a distribution both spatially and in size population; as the critically strained regions may only occupy a small area of the sheet, there is also a probabilistic aspect. The critical region may, or may not contain a large defect and therefore there is likely to be some scatter in measured limit strains and the forming limit curve is more properly a region of increasing probability of failure.

5.5.5 Anisotropy

The shape of the yield locus is shown to influence the forming limit in biaxial tension. This locus changes if the material becomes anisotropic. If a quadratic yield function is used, as in Equation 2.11, anisotropy in the sheet, characterized by an R-value >1, will cause the locus to be extended along the biaxial stress axis as shown in Figure 5.21(a). The effect of this in a numerical analysis of the forming limit would be to reduce the biaxial strain limit. This is not observed experimentally and it appears that a different yield function employing higher exponents, 6–8, is more realistic for certain materials. The shape of a yield locus for a high exponent law is shown in Figure 5.21(b) and for such a model it

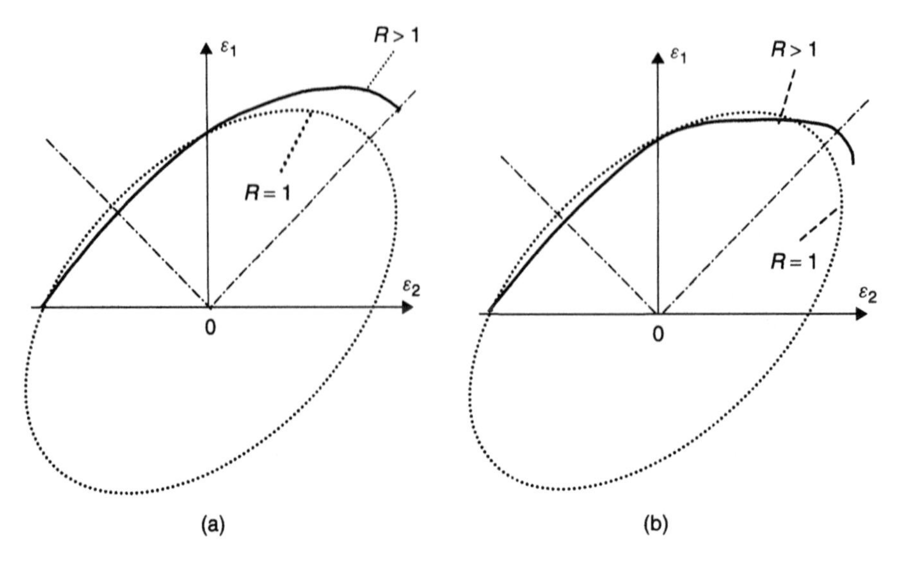

Figure 5.21 Plane stress yield loci. The dotted ellipses are for a quadratic function for an isotropic material; the bold lines are for a high R-value material with (a) a quadratic yield function and (b) a high exponent function.

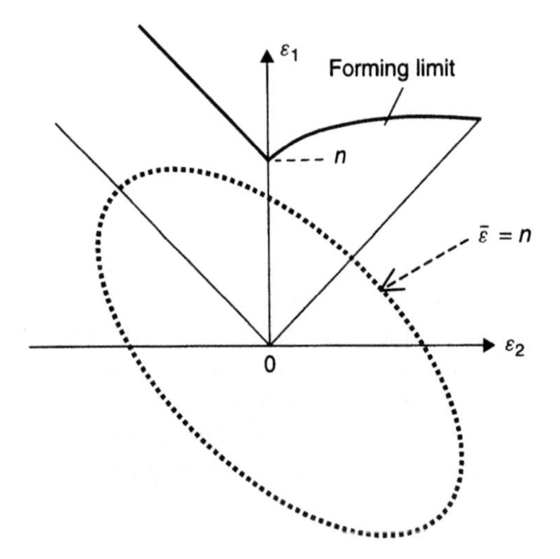

Figure 5.22 Diagram showing the strain envelope in which material data can be obtained from the tensile test.

has been shown that changes in the R-value do not have a significant influence on the forming limit curve.

5.5.6 Other considerations

In the tensile test, stress strain data can only be obtained up to the onset of diffuse necking, i.e. up to an effective strain $\bar{\varepsilon} \leq n$. The envelope of this strain is shown in Figure 5.22

and it may be seen that local necking occurs at strains rather greater than this. In many analyses, the tensile data is extrapolated, assuming that the strain-hardening index remains the same at high strains. This may not be the case and caution should be exercised.

There may also be an interaction between material properties in the way that they influence the forming limit curve. In Figure 5.23, an example is given in which increasing the strain-hardening index may not increase the forming limits in all forming paths, if the change in, n, is an accompanied by a reduction in the fracture strain. As seen here, fracture will reduce the forming limit in equal biaxial tension, even though the forming limit is increased in other regions by an increase in the strain-hardening index. In the lubricated Olsen and Erichsen tests in which sheet is stretched over a well-lubricated hemispherical punch, failure occurs nearly in biaxial tension; in comparing some materials, differences may be due to different fracture properties rather than differences in strain-hardening.

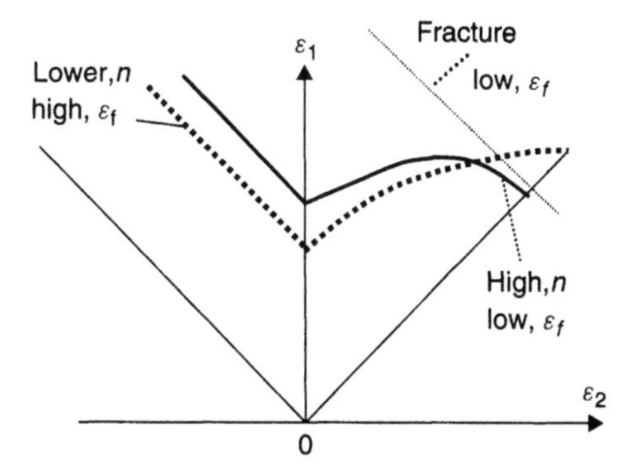

Figure 5.23 Diagram showing the changes in the forming limit curve when there is a property change such that strain-hardening is increased and the fracture strain lowered.

5.6 The forming window

In summary, sheet metal forming processes can be limited by various events. In the preceding sections, failure by a process of local necking and tearing has been examined. It is also possible for ductile sheet to fracture either within a necked region or before necking is established. Other limitations include wrinkling of the sheet under compressive loading. We have seen also that, for practical reasons, sheet can only be deformed by tensile forces and therefore one of the principal stresses must be positive, or, in the limit equal to zero. Taking these factors into account, it is useful to identify a *forming window* in which plane stress sheet forming is possible. This is illustrated in Figure 5.24.

The compressive limit where the major tension just reduces to zero is shown at the strain path of $\beta = -2$. The wrinkling limit is not solely a material property and therefore the limit is only shown as a region in the second and fourth quadrants. The diagram is a pictorial aid and it can be seen that if the strain-hardening index becomes small, the window shrinks to a narrow slit along the left-hand diagonal. As mentioned, strengthening processes usually lead to reduced strain-hardening in the sheet and one of the challenges of sheet metal forming is to devise processes for forming strong materials that will permit safe straining even though the window is small.

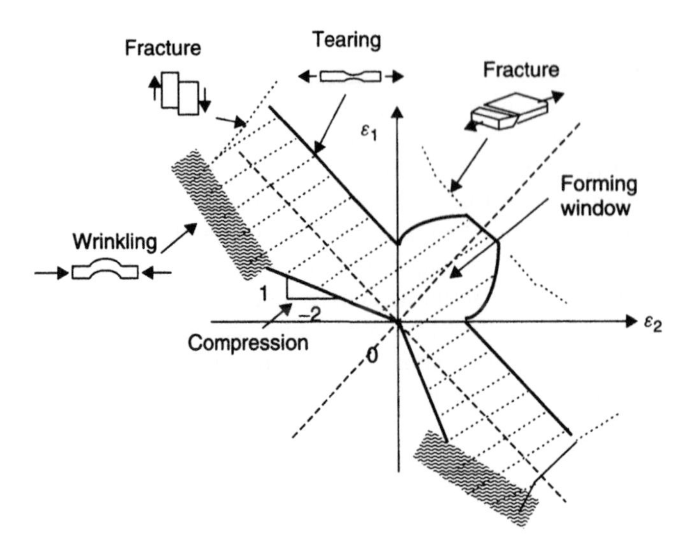

Figure 5.24 The forming window for plane stress forming of sheet.

Note: The approximate solution for Equation 5.12 is arrived at as follows. The equation can be written as

$$\left[1-\frac{n-\varepsilon_u}{n}\right]\exp\left(\frac{n-\varepsilon_u}{n}\right)=\left(1+\frac{dA_0}{A_0}\right)^{\frac{1}{n}}$$

As indicated $(n-\varepsilon_u)$ and dA_0/A_0 are small compared with unity and the above equation may be approximated as

$$\left(1-\frac{n-\varepsilon_u}{n}\right)\left(1+\frac{n-\varepsilon_u}{n}\right)=1-\left(\frac{n-\varepsilon_u}{n}\right)^2\approx 1+\frac{dA_0}{A_0}\frac{1}{n}$$

This gives Equation 5.13.

5.7 Exercises

Ex. 5.1 Cold-rolled steel obeys the law $\sigma_f = K(\varepsilon_0+\varepsilon)^n$.

(a) Determine the strain at which the maximum load is reached in a uniform tensile strip.
(b) What happens when $\varepsilon_0 > n$?

[Ans: (a) $n-\varepsilon_0$; (b) $\varepsilon_1^* = 0$]

Ex. 5.2 Figure 5.25 shows a 100 mm length of a tensile test-piece in which 10 mm has a width of 12.4 mm and the remainder 12.5 mm. The thickness is uniform at the start, $t_0 =$ 1.2 mm. The material obeys an effective stress strain law $\overline{\sigma} = 750\overline{\varepsilon}^{0.22}$ MPa. Assuming that each length deforms in uniaxial tension, determine the maximum load and the final

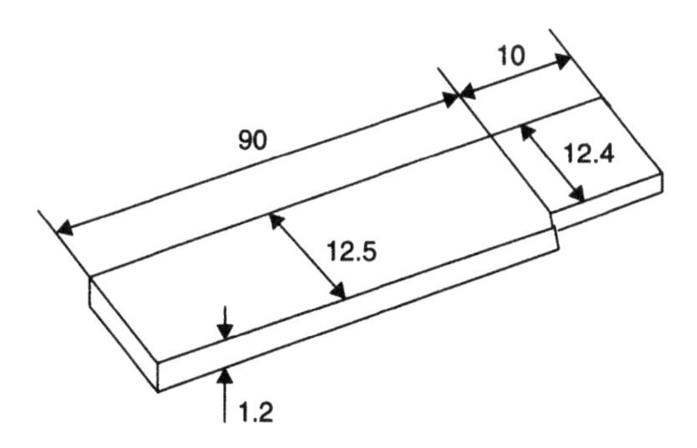

Figure 5.25 Dimensions of test-piece for Exercise 5.2.

length of a 20 mm gauge length in the wider section and the maximum strain in this section.
[Ans: $P_{max} = 6.42$ KN, $\varepsilon_{1A} = 0.17$, $l = 23.7$ mm]

Ex. 5.3 A method is proposed for measuring the strain-hardening index in sheet as defined in Section 1.1.3. A test-piece is used that has two parallel reduced lengths, one is 10.0 mm width and the other 9.8 mm width. In the wider section a gauge length of 50 mm is marked. The strip is pulled to failure and the gauge length measured to determine the true strain ε_a. Obtain a diagram relating the true strain ε_a to the strain-hardening index n for the range $0.05 < \varepsilon_a < 0.2$.

Ex. 5.4 A strip subjected to tension consists of two regions of equal length l, one of cross-sectional area A_a the other A_b. The material is perfectly plastic but is rate sensitive so that the effective stress strain rate law is $\overline{\sigma} = B\,(\dot{\varepsilon}_{eff.})^m$. If the extension rate of the combined strip is v, determine the strain rate in each section, $\dot{\varepsilon}_1$and $\dot{\varepsilon}_2$.

$$\left[Ans: \frac{v}{l\left[1 + (A_a/A_b)^{1/m}\right]}; \frac{v}{l\left[1 + (A_b/A_a)^{1/m}\right]}\right]$$

Ex. 5.5 An element of material has an imperfection characterized by $f_0 = 0.995$ as shown in Figure 5.12. It is deformed in equal biaxial tension, $\sigma_{1a} = \sigma_{2a}$. The material obeys an effective stress strain law $\overline{\sigma} = 600\,(0.004 + \overline{\varepsilon})^{0.2}$ MPa. Determine the principal stresses and the stress ratio in the groove when the uniform region starts to deform.
[Ans: 199.9, 197.9, 0.990]

6 Bending of sheet

6.1 Introduction

Bending along a straight line is the most common of all sheet forming processes; it can be done in various ways such as forming along the complete bend in a die, or by wiping, folding or flanging in special machines, or sliding the sheet over a radius in a die. A very large amount of sheet is roll formed where it is bent progressively under shaped rolls. Failure by splitting during a bending process is usually limited to high-strength, less ductile sheet and a more common cause of unsatisfactory bending is lack of dimensional control in terms of springback and thinning.

If the line of bending is curved, adjacent sheet is usually deformed in the process and the sheet is either stretched, which may lead to splitting, or compressed with the possibility of buckling. There are special cases where sheet can be bent along curved lines without stretching or shrinking adjacent areas, but these require special geometric design.

In this chapter, simple cases of bending along straight lines are examined for the elastic, plastic and fully plastic regimes.

6.2 Variables in bending a continuous sheet

As shown in Figure 6.1, we consider a unit width of a continuous sheet in which a cylindrical bent region of radius of curvature ρ is flanked by flat sheet. The bend angle is θ, and a *moment per unit width* M,and a tension (*force per unit width*) T are applied. We note that the tension T is applied at the middle surface of the sheet. The units of M are [force][length]/[length] and of T [force]/[length].

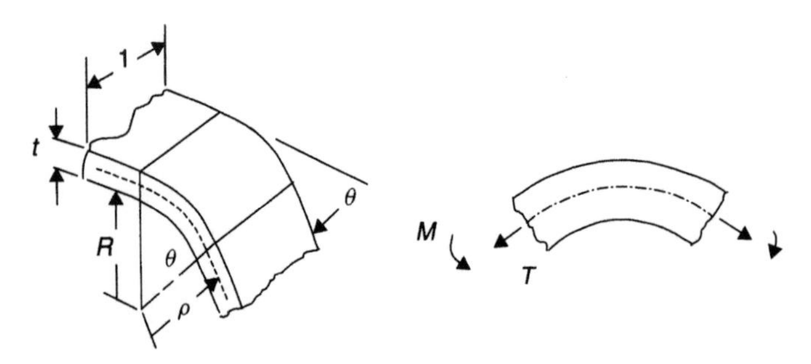

Figure 6.1 A unit length of a continuous strip bent along a line.

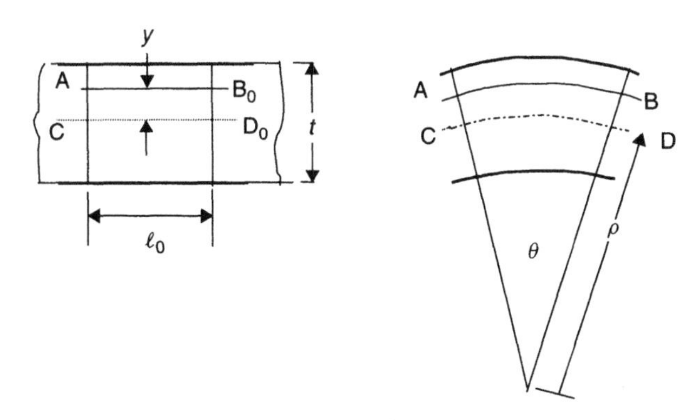

Figure 6.2 Deformation of longitudinal fibres in bending and tension.

6.2.1 Geometry and strain in bending

In bending a thin sheet to a bend radius more than three or four times the sheet thickness; it may be assumed that a plane normal section in the sheet will remain plane and normal and converge on the centre of curvature as shown in Figure 6.2.

In general, a line CD_0 at the middle surface may change its length to CD if, for example, the sheet is stretched during bending; i.e. the original length l_0 becomes

$$l_s = \rho\theta \tag{6.1}$$

A line AB_0 at a distance y from the middle surface will deform to a length

$$l = \theta\,(\rho + y) = \rho\theta\left(1 + \frac{y}{\rho}\right) = l_s\left(1 + \frac{y}{\rho}\right) \tag{6.2}$$

The axial strain of the fibre AB is

$$\varepsilon_1 = \ln\frac{l}{l_0} = \ln\frac{l_s}{l_0} + \ln\left(1 + \frac{y}{\rho}\right) = \varepsilon_a + \varepsilon_b \tag{6.3}$$

where ε_a is the strain at the middle surface or the membrane strain and ε_b is the bending strain. Where the radius of curvature is large compared with the thickness, the bending strain can be approximated as,

$$\varepsilon_b = \ln\left(1 + \frac{y}{\rho}\right) \approx \frac{y}{\rho} \tag{6.4}$$

The strain distribution is approximately linear as illustrated in Figure 6.3.

6.2.2 Plane strain bending

If the flat sheet on either side of the bend in Figure 6.1 is not deforming it will constrain the material in the bend to deform in plane strain; i.e. the strain parallel to the bend will be zero. In this work, plane strain conditions will be assumed, unless stated otherwise. The deformation process in bending an isotropic sheet is therefore

$$\varepsilon_1;\quad \varepsilon_2 = 0;\quad \varepsilon_3 = -\varepsilon_1$$
$$\sigma_1;\quad \sigma_2 = \sigma_1/2;\quad \sigma_3 = 0 \tag{6.5}$$

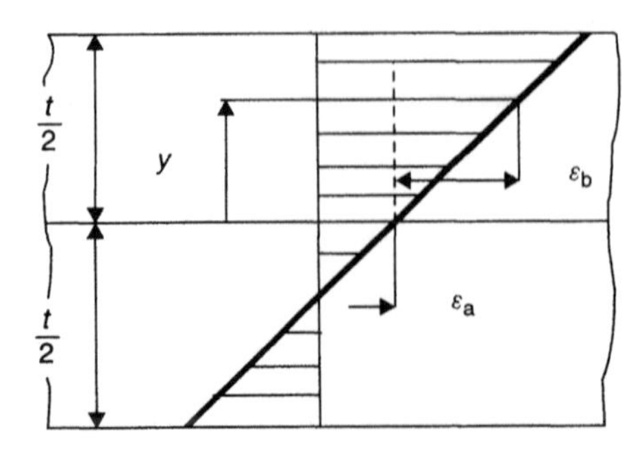

Figure 6.3 Assumed strain distribution in bending.

Following Equations 2.18(b) and 2.19(c), for, $\beta = 0, \alpha = 1/2$, we obtain

$$\sigma_1 = \frac{2}{\sqrt{3}}\sigma_f = S \quad \text{and} \quad \varepsilon_1 = \frac{\sqrt{3}}{2}\bar{\varepsilon} \tag{6.6}$$

where S is the *plane strain flow stress*. (Equation 6.6 assumes the von Mises yield condition. If a Tresca yield criterion is assumed, $\sigma_1 = \sigma_f = S$.) The stresses on a section along the bend axis are illustrated in Figure 6.4. Clearly, at the edge of the sheet, the stress along the bend axis will be zero at the free surface and plane strain will not exist. It is usually observed that the edge of the sheet will curl as illustrated. This happens because the stress state is approximately uniaxial tension near the edges of the sheet; the minor strain will be negative near the outer surface and positive near the inner surface giving rise to the *anticlastic* curvature as shown. Within the bulk of the sheet, however, plane strain deformation is assumed with the minor strain along the axis of the bend equal to zero.

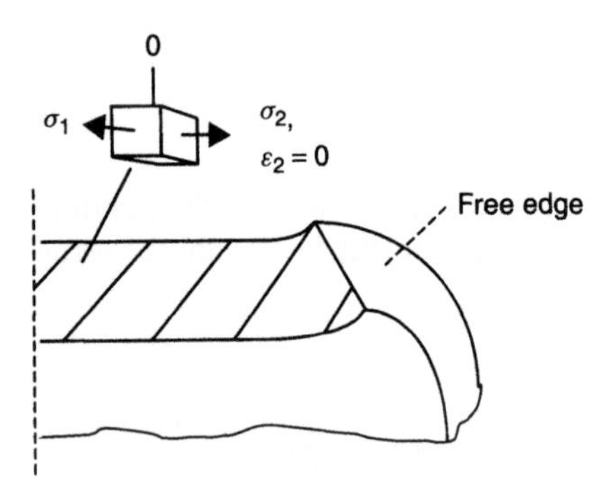

Figure 6.4 Stress state on a section through the sheet in plane strain bending.

6.3 Equilibrium conditions

We consider a general stress distribution on a normal section through a unit width of sheet in bending, as shown in Figure 6.5. The force acting on a strip of thickness dy across the unit section is $\sigma_1 \times dy \times 1$. The tension T on the section is in equilibrium with the integral of this force element, i.e.

$$T = \int_{-t/2}^{t/2} \sigma_1 \, dy \tag{6.7}$$

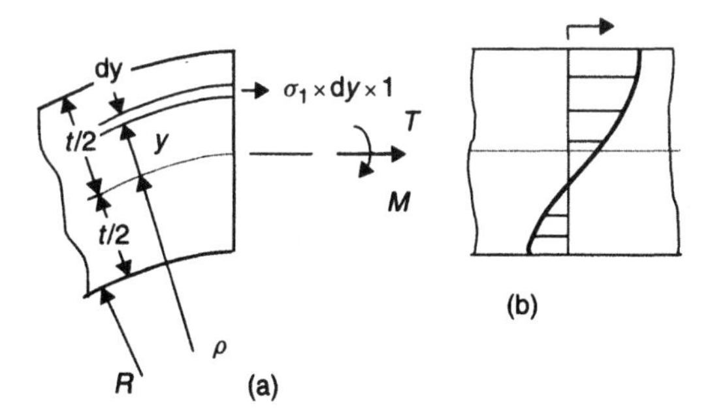

Figure 6.5 Equilibrium diagram (a) for a section through a unit width of sheet and (b) a typical stress distribution.

Integrating the moment of the force element, we obtain

$$M = \int_{-t/2}^{t/2} \sigma_1 \, dy 1 y = \int_{-t/2}^{t/2} \sigma_1 y \, dy \tag{6.8}$$

(We note too that there is a third equilibrium equation for forces in the radial direction arising from the tension T. This is given in Section 4.2.5 by Equation 4.11.)

6.4 Choice of material model

For the strain distribution given by Equation 6.3, the stress distribution on a section can be determined if a stress strain law is available. In general, the material will have an elastic, plastic strain-hardening behaviour as shown in Figure 6.6(a). In many cases, it is useful to approximate this by a simple law and several examples will be given. The choice of material model will depend on the magnitude of the strain in the process. The strain will depend mainly on the *bend ratio*, which is defined as the ratio of the radius of curvature to sheet thickness, ρ/t.

6.4.1 Elastic, perfectly plastic model

If the bend ratio is not less than about 50, strain-hardening may not be so important and the material model can be that shown in Figure 6.6(b). This has two parts, i.e. if the stress

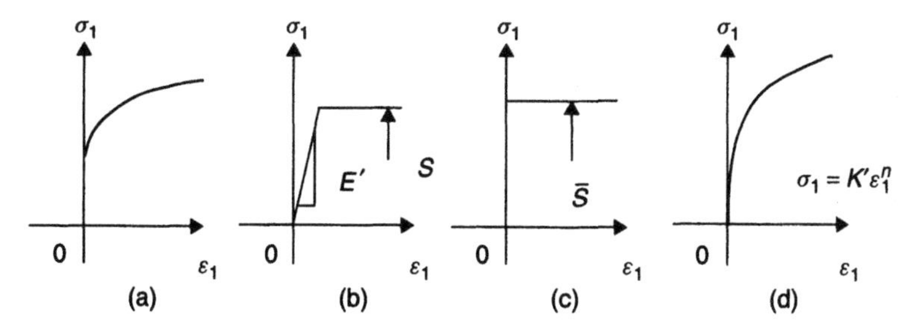

Figure 6.6 Material models for bending. (a) An actual stress–strain curve. (b) An elastic, perfectly plastic model. (c) A rigid, perfectly plastic model. (d) A strain-hardening plastic model.

is less than the plane strain yield stress, S

$$\sigma_1 = E'\varepsilon_1 \tag{6.9}$$

where the modulus of elasticity in plane strain is slightly different from the uniaxial Young's modulus, E; i.e.

$$E' = \frac{E}{1-\nu^2} \tag{6.10}$$

where ν is Poisson's ratio.

For strains greater than the yield strain,

$$\sigma_1 = S \tag{6.11}$$

where S is constant. In isotropic materials, S is related to the uniaxial flow stress by Equation 6.6 for the von Mises yield condition.

6.4.2 Rigid, perfectly plastic model

For smaller radius bends, and where we are not concerned with elastic springback, it may be sufficient to neglect both elastic strains and strain-hardening. A rigid, perfectly plastic model is shown in Figure 6.6(c), where

$$\sigma_1 = \overline{S} \tag{6.12}$$

and $\overline{S}$ is a value averaged over the strain range as indicated in Section 3.5.4.

6.4.3 Strain-hardening model

Where the strains are large, the elastic strains may be neglected and the power law strain-hardening model used, where

$$\sigma_1 = K'\varepsilon_1^n \tag{6.13}$$

For a material having a known effective stress–strain curve of the form

$$\overline{\sigma} = \sigma_f = K\overline{\varepsilon}^n \tag{6.14}$$

the strength coefficient K' can be calculated using Equations 6.6. This model is illustrated in Figure 6.6(d).

6.5 Bending without tension

Where sheet is bent by a pure moment without any tension being applied, the neutral axis will be at the mid-thickness. This kind of bending is examined here for several types of material behaviour. In these cases, a linear strain distribution as illustrated in Figure 6.3 is assumed and the equilibrium equations, Equations 6.7 and 6.8, will apply.

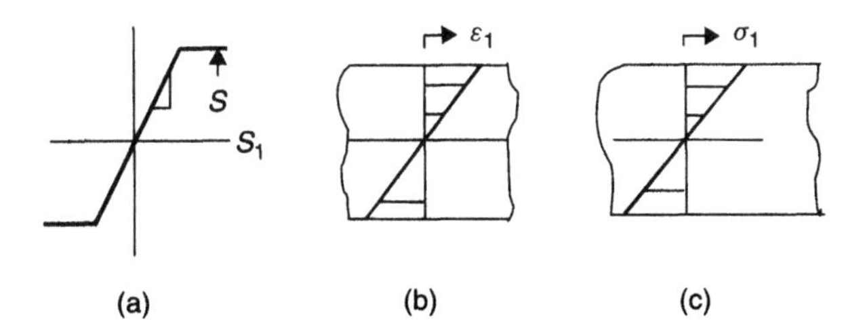

Figure 6.7 Linear elastic bending of sheet showing the material model (a), the strain distribution (b), and the stress distribution (c).

6.5.1 Elastic bending

The material model is illustrated in Figure 6.7(a) where the yield stress is S. The stress–strain relation is given by Equation 6.9 and for the strain distribution shown in Figure 6.7(b), the stress distribution in Figure 6.7(c) will be obtained.

The stress at a distance y from the neutral axis, from Equations 6.4 and 6.13, is

$$\sigma_1 = E'\varepsilon_b = E'\frac{y}{\rho} \tag{6.15}$$

The moment at the section, from Equation 6.8, is

$$M = \int_{-t/2}^{t/2} E'\frac{y}{\rho} y\,\mathrm{d}y = 2\frac{E'}{\rho}\int_0^{t/2} y^2\,\mathrm{d}y = \frac{E'}{\rho}\frac{t^3}{12} \tag{6.16}$$

From Equation 6.15, we have

$$\frac{E'}{\rho} = \frac{\sigma_1}{y}$$

and Equation 6.16 can be written in the familiar form for elastic bending:

$$\frac{M}{I} = \frac{\sigma_1}{y} = E'\left(\frac{1}{\rho}\right) \tag{6.17}$$

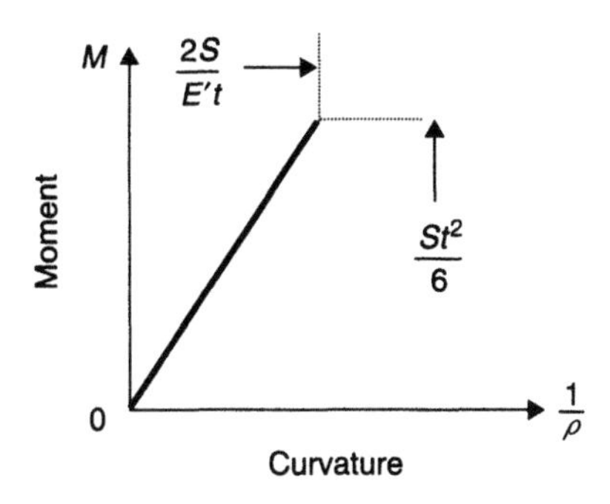

Figure 6.8 Moment curvature diagram for elastic bending.

where $I = t^3/12$ is the second moment of area for a unit width of sheet and $1/\rho$ is the curvature.

The limit of elastic bending is when the outer fibre at $y = t/2$ reaches the plane strain yield stress S. The limiting elastic moment is given by

$$M_e = \frac{St^2}{6} \tag{6.18}$$

and the curvature at this moment is

$$\left(\frac{1}{\rho}\right)_e = \frac{2S}{E't} \tag{6.19}$$

Within this elastic range, the moment, curvature diagram is linear as shown in Figure 6.8, i.e.

$$M = \frac{E't^3}{12}\left(\frac{1}{\rho}\right) \tag{6.20}$$

The bending stiffness of unit width of the sheet is $E't^3/12$.

6.5.2 Rigid, perfectly plastic bending

If the curvature is greater than about five times the limiting elastic curvature, a rigid, perfectly plastic model, Equation 6.12, as shown in Figure 6.6(c), may be appropriate, although this will not give information on springback. The stress distribution will be as shown in Figure 6.9. In Equation 6.8, the stress is constant and integrating, we obtain the so-called *fully plastic moment* M_p as

$$M_p = \frac{St^2}{4} \tag{6.21}$$

The moment will remain constant during bending and is illustrated in Figure 6.10.

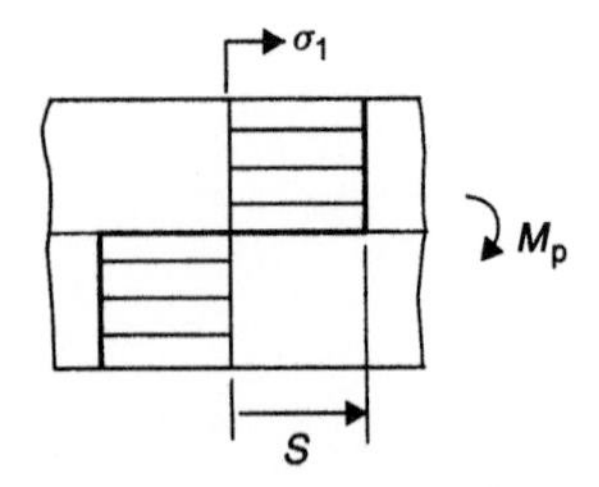

Figure 6.9 Stress distribution for a rigid, perfectly plastic material bent without tension.

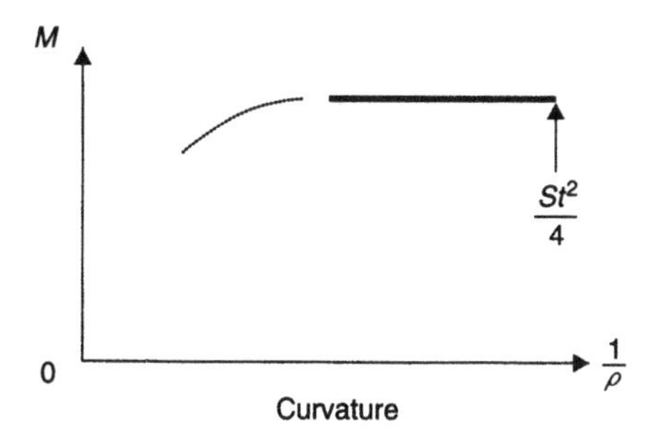

Figure 6.10 The moment curvature diagram for a rigid, perfectly plastic sheet bent without tension.

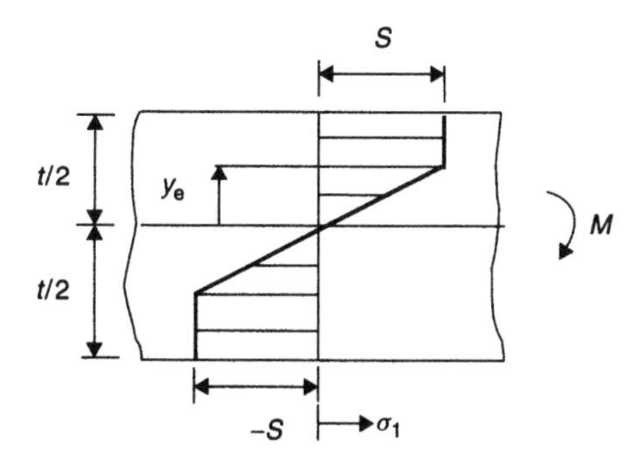

Figure 6.11 Stress distribution for an elastic, perfectly plastic sheet bent without tension.

6.5.3 Elastic, perfectly plastic bending

For curvatures beyond the limiting elastic curvature $(1/\rho)_e$ and below that where the moment reaches the fully plastic moment M_p, an elastic, perfectly plastic model, as in Section 6.4.1, is often used. The model is illustrated in Figure 6.6(b); the flow stress is constant and for plane strain, $\sigma_1 = \left(2/\sqrt{3}\right)\sigma_f = S$. The stress distribution is illustrated in Figure 6.11; for $y > y_e$, the material is plastic with a flow stress S. As the curvature increases, y_e decreases and at any instant is given by

$$(\varepsilon_b)_{y=y_e} = \frac{y}{\rho} = \frac{S}{E'} \quad \text{i.e.} \quad y_e = \frac{S}{E'}\frac{1}{(1/\rho)} = m\frac{t}{2} \tag{6.22}$$

From Equation 6.19,

$$m = \frac{(1/\rho)e}{(1/\rho)}$$

and $1 \geq m \geq 0$.
The equilibrium equation, from Equation 6.16, is

$$M = 2\left\{\int_0^{y_e} E'\frac{y}{\rho}y\,dy + \int_{y_e}^{t/2} Sy\,dy\right\} = \frac{St^2}{12}(3 - m^2) \tag{6.23}$$

The moment, curvature characteristic is shown in Figure 6.12 and it may be seen that this is tangent to the elastic curve at the one end and to the fully plastic curve at the other.

It may be seen that with this non-strain-hardening model, the moment still increases beyond the limiting elastic moment and reaches $1.5M_e$ before becoming constant. For this reason, elastic plastic bending is usually a stable process in which the curvature increases uniformly in the sheet without kinking. It may be shown that for materials that do not fit this elastic, perfectly plastic model, for example aged sheet having a stress–strain curve as shown in Figure 1.4, the moment characteristic is different and kinking may occur.

In real materials it is very difficult to predict precisely the moment curvature characteristic in the region covered by the bold curve in Figure 6.12 from tensile data. The moment characteristic is extremely sensitive to material properties at very small strain and these properties often are not determined accurately in a tension test.

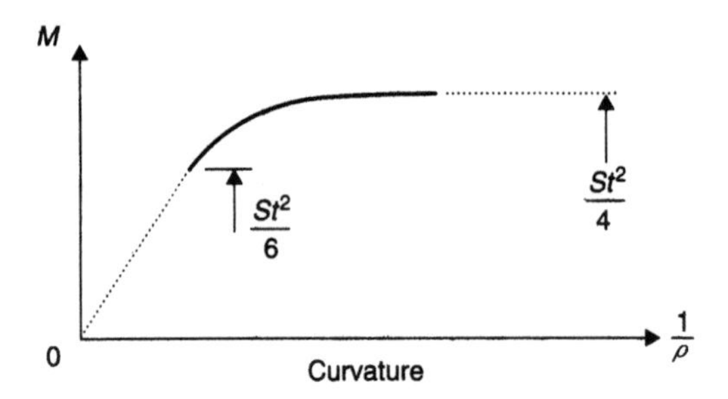

Figure 6.12 Moment curvature diagram for an elastic, perfectly plastic sheet bent without tension.

6.5.4 Bending of a strain-hardening sheet

If a power law strain-hardening model of the kind shown in Section 6.4.3 and Figure 6.6(d) is used, the stress distribution will be as shown in Figure 6.13. The whole section is assumed to be deforming plastically and the stress at some distance, y, from the middle surface is

$$\sigma_1 = K'\varepsilon_1^n \approx K'\left(\frac{y}{\rho}\right)^n \tag{6.24}$$

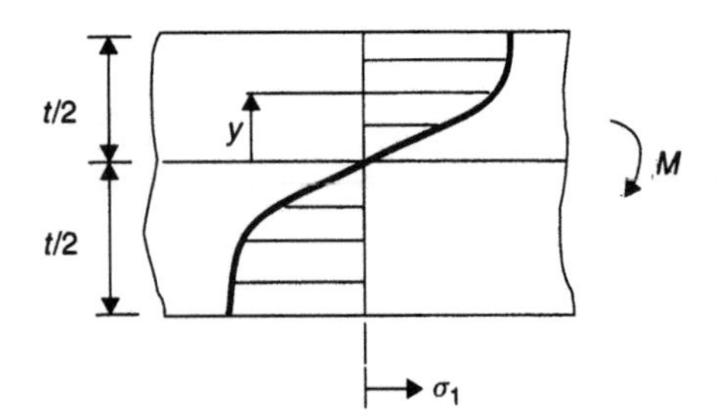

Figure 6.13 Stress distribution for a power-law-hardening sheet bent without tension.

The equilibrium equation can be written as

$$M = 2K'\left(\frac{1}{\rho}\right)^n \int_0^{t/2} y^{1+n}\,\mathrm{d}y = K'\left(\frac{1}{\rho}\right)^n \frac{t^{n+2}}{(n+2)\,2^{n+1}} \tag{6.25}$$

These equations can be combined to give a set of equations for bending a non-linear material, i.e.

$$\frac{M}{I_n} = \frac{\sigma_1}{y^n} = K'\left(\frac{1}{\rho}\right)^n \tag{6.26}$$

where

$$I_n = \frac{t^{n+2}}{2^{n+1}(n+2)} \tag{6.27}$$

The moment, curvature diagram is shown in Figure 6.14.

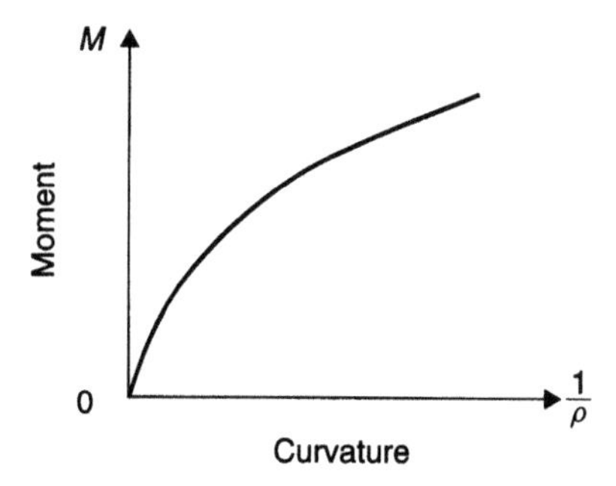

Figure 6.14 Moment curvature diagram for a strain-hardening sheet bent without tension.

The power law equation is not a good fit for most materials at very small strains so that Equation 6.26 will not predict the moment curvature characteristic near the elastic, plastic transition

Equations 6.26 reduces to some familiar relations for special cases. For the linear elastic model, $n = 1$, $K' = E'$, and we obtain Equations 6.17. For the rigid, perfectly plastic model, $n = 0$, $K' = \overline{S}$, and Equation 6.26 reduces to Equation 6.21.

6.5.5 (Worked example) moments

A hard temper aluminium sheet, 2 mm thick, has a flow stress of 120 MPa, that is approximately constant. Determine the moment per unit width to bend the sheet to the limiting elastic state. What is the radius of curvature at this moment? Determine the fully plastic moment if the sheet is bent further. The elastic modulus is 70 GPa and Poisson's ratio is 0.3.

Solution

From Equation 6.6, the plane strain plastic bending stress is

$$S = \frac{2}{\sqrt{3}}\sigma_f = 138.6\,\text{MPa}$$

From Equation 6.10, the elastic modulus in plane strain is

$$E' = \frac{E}{1-\upsilon^2} = 76.9\,\text{GPa}$$

From Equation 6.18, the limiting elastic bending moment is

$$M_e = \frac{St^2}{6} = \frac{138.6 \times 10^6 \left(2 \times 10^{-3}\right)^2}{6} = 92.3\,\text{Nm/m}$$

The radius of curvature, from Equation 6.19, is

$$\rho_e = \frac{E't}{2S} = \frac{76.9 \times 10^9 \times 2 \times 10^{-3}}{2 \times 138.6 \times 10^6} = 0.555\,\text{m or } 555\,\text{mm}$$

The fully plastic moment from Equation 6.21 is

$$M_p = \frac{St^2}{4} = \frac{3}{2}M_e = 138\,\text{Nm/m}$$

6.6 Elastic unloading and springback

If a sheet is bent by a moment to a particular curvature, as shown in Figure 6.15, and the moment then released, there will be a change in curvature and bend angle. The length of the mid-surface is

$$l = \rho\theta$$

This will remain unchanged during unloading as the stress and strain at the middle surface are zero. From this, we obtain

$$\theta = l\frac{1}{\rho} \tag{6.28}$$

Differentiating Equation 6.28, in which l = constant, we obtain

$$\frac{\Delta\theta}{\theta} = \frac{\Delta(1/\rho)}{1/\rho} \tag{6.29}$$

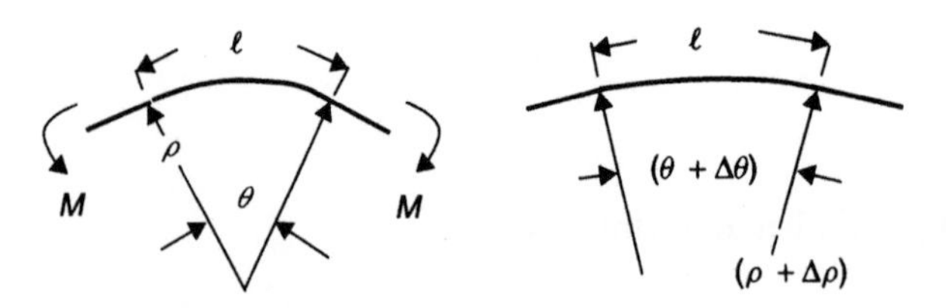

Figure 6.15 Unloading a sheet that has been bent by a moment without tension.

6.6.1 Springback in an elastic, perfectly plastic material

The assumed stress–strain curve for an elastic, perfectly plastic material that undergoes reverse loading is shown in Figure 6.16. (This neglects any Bauschinger effect; this is the phenomenon of softening on reverse loading that is observed in many materials.) From Figure 6.16, a change in stress of $\Delta\sigma_1 = -2S$ can occur without the material becoming plastic.

If we assume that the unloading of the sheet will be an elastic process, then the elastic bending equations, Equations 6.17, can be written in difference form, i.e.

$$\frac{\Delta M}{I} = \frac{\Delta\sigma_1}{y} = \frac{\Delta\sigma_{1\,\max}}{t/2} = E'\Delta\left(\frac{1}{\rho}\right) \tag{6.30}$$

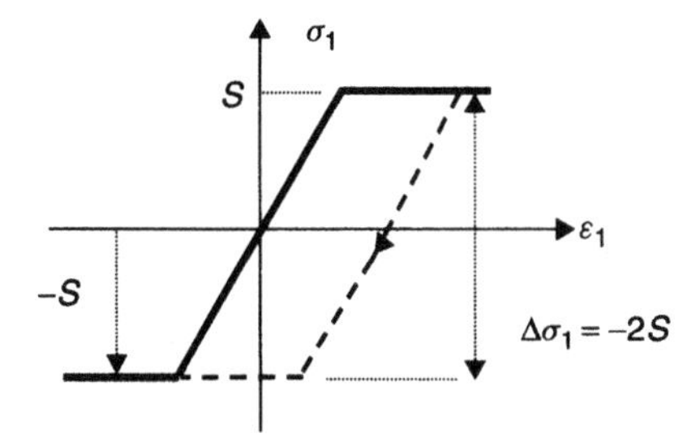

Figure 6.16 Elastic, perfectly plastic material model with reverse loading.

For a sheet that has been bent to the fully plastic moment, the unloading curve will be parallel to the elastic loading line as shown in Figure 6.17. Noting the similar triangles, we see that for a change in moment of $-M_p$,

$$\frac{\Delta(1/\rho)}{(1/\rho)_e} = \frac{\Delta M}{M_e} = \frac{-M_p}{M_e}$$

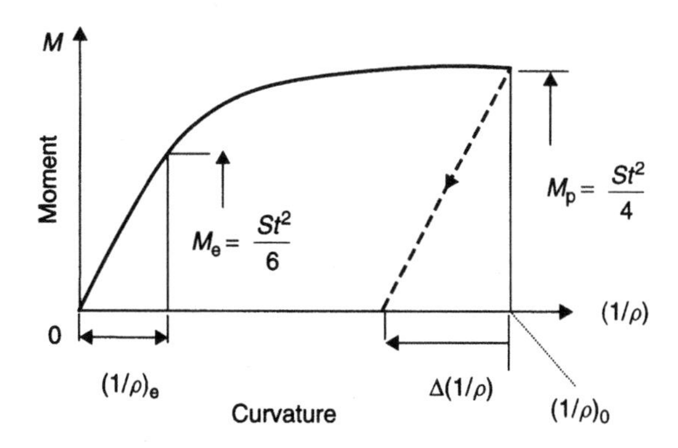

Figure 6.17 Moment, curvature diagram for an elastic, perfectly plastic sheet showing unloading from a fully plastic moment.

The ratio of the fully plastic moment to the limiting elastic moment has been shown to be

$$\frac{M_p}{M_e} = \frac{3}{2}$$

Therefore combining the above with Equation 6.19, we obtain

$$\Delta\left(\frac{1}{\rho}\right) = -\frac{3}{2}\left(\frac{1}{\rho}\right)_e = -3\frac{S}{E't} \tag{6.31}$$

If the sheet has been unloaded from a curvature of $(1/\rho)_0$, the proportional change in curvature, from Equation 6.31 is

$$\frac{\Delta\,(1/\rho)}{(1/\rho)_0} = -3\frac{S}{E'}\frac{\rho_0}{t} \tag{6.32}$$

or, from Equation 6.29, the change in bend angle is

$$\Delta\theta \approx -3\frac{S}{E'}\frac{\rho_0}{t}\theta \tag{6.33}$$

Equation 6.33 is only approximate and applies to small differences in angle or curvature and to the case in which the sheet has been bent to a nearly fully plastic state.

Nevertheless, the equation is very useful and indicates that springback is proportional to:

- the ratio of flow stress to elastic modulus, S/E', which is small and often of the order of 1/1000;
- the bend ratio ρ_0/t;
- the bend angle.

Thus springback will be large when thin high strength sheet is bent to a gentle curvature.

6.6.2 Residual stresses after unloading

When an elastic, perfectly plastic sheet is unloaded from a fully plastic state, it is shown above that the change in moment is $\Delta M = -M_p$. Substituting in Equation 6.30

$$\frac{-St^2/4}{t^3/12} = \frac{\Delta\sigma_{1\,\max}}{t/2}$$

i.e. the change in stress at the outer fibre is

$$\Delta\sigma_1 = -\frac{3}{2}S \tag{6.34}$$

(Equation 6.34 supports the assumption that for the simple bending model given here, the unloading process is fully elastic.)

Thus the effect of unloading is equivalent to adding an elastic stress distribution of maximum value of −3S/2 to the fully plastic stress state as shown in Figure 6.18. The residual stress distribution is shown on the right of Figure 6.18; this an is idealized representation arising from the simple model, but it does show that after unloading, the tension

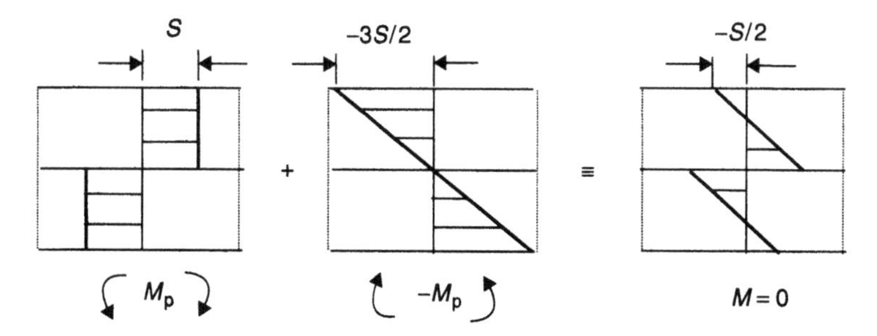

Figure 6.18 Residual stress distribution after unloading from a fully plastic moment.

side of the bend would have a significant compressive residual stress at the surface and there would be a residual tensile stress on the inner surface.

6.6.3 Reverse bending

If a sheet has been bent to a fully plastic state and unloaded, it is interesting to see what reverse bending is required to cause renewed plastic deformation. From Figure 6.16, it may be seen that the change in stress required at the outer fibre to just start yielding is $-2S$. Substituting in Equation 6.30, shows that the change in moment is

$$\Delta M = -\frac{2S}{t/2}\frac{t^3}{12} = -\frac{St^2}{3} \tag{6.35}$$

The moment for reverse yielding is therefore

$$M_{\text{rev.}} = St^2\left(\frac{1}{4} - \frac{1}{3}\right) = -\frac{St^2}{12} = -\frac{M_e}{2} \tag{6.36}$$

Thus yielding starts at only half the initial yield moment as shown in Figure 6.19. This softening effect is important as there are a number of processes in which sheet goes through bend–unbend and reverse bend cycles. The actual softening is likely to be greater than that calculated above as most materials will also have some Bauschinger effect and yield at a reverse stress of magnitude less than S.

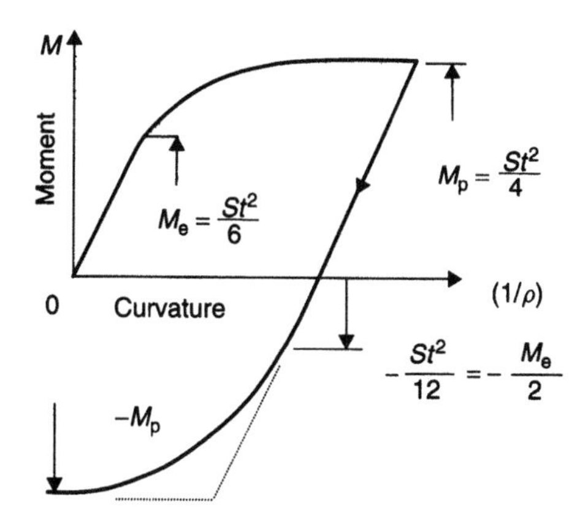

Figure 6.19 Reverse bending of an elastic, perfectly plastic sheet.

6.7 Small radius bends

6.7.1 Strain distribution

In the previous sections, the strain in bending was assumed to be a linear function of the distance from the middle surface. If the radius of the bend is approximately the same as the sheet thickness, a more refined analysis is necessary. In Figure 6.20, a length l_0 of sheet of thickness t is bent under plane strain and constant thickness conditions to a middle surface radius of ρ. In the deformed shape, the length of the middle surface is

$$l_a = \rho\theta$$

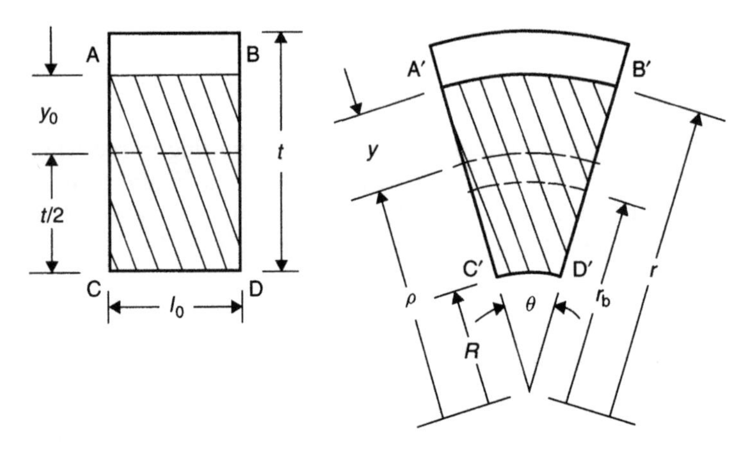

Figure 6.20 Element of sheet bent to a small radius bend.

and the volume of the deformed element is,

$$\frac{\theta}{2}\left(R_0^2 - R^2\right)1$$

where $R_0 = \rho + (t/2)$ is the radius of the outer surface and $R = \rho - (t/2)$ is the radius of the inner surface. As the volume of the element remains constant, we obtain from the above

$$l_0 t 1 = \frac{\theta}{2}\left(R_0^2 - R^2\right)1 = \theta\rho t$$

i.e.

$$l_0 = \theta\rho \tag{6.37}$$

If the length of the middle surface in the deformed condition is l_a then,

$$l_a = \rho\theta = l_0 \tag{6.38}$$

i.e. the length of the middle surface does not change.

In the previous simple analysis, it was assumed that a fibre at some distance y from the middle surface remained at that distance during bending. In small radius bends, this is not the case. We consider a fibre y_0 from the middle surface in the undeformed state as shown in Figure 6.20. Equating the shaded volumes shown,

$$\frac{\theta}{2}\left\{r^2-\left[\rho-\frac{t}{2}\right]^2\right\}=\left(y_0+\frac{t}{2}\right)l_0 \tag{6.39}$$

Given that, $\theta = l/\rho = l_0/\rho$, the radius of the fibre initially at y_0 is

$$r=\sqrt{2\rho y_0+\rho^2+\frac{t_2}{4}} \tag{6.40}$$

The length of the fibre, A′B′, is $l = r\theta$, and substituting $\theta = l_0/\rho$ we obtain

$$\frac{l}{l_0}=\sqrt{1+\frac{2y_0}{\rho}+\left(\frac{t_0}{2\rho}\right)^2} \tag{6.41}$$

The change in fibre length on bending is shown in Figure 6.21(a). Fibres initially above the middle surface will always increase in length. Fibres below the middle surface will decrease in length initially, but may then increase. The minimum length is denoted by the point B in Figures 6.21(a) and (b). The minimum length is found by differentiating Equation 6.41 with respect to curvature $1/\rho$ and equating to zero. From this, the strain in a fibre begins to reverse when

$$\frac{t}{\rho}=-\frac{4y_0}{t} \tag{6.42}$$

Substituting in Equation 6.41, the minimum length of such a fibre is

$$\frac{l}{l_0}=\sqrt{1-\left(\frac{2y_0}{t}\right)^2} \tag{6.43}$$

and the radius of the fibre at B that has reached its minimum length is

$$r_b=\sqrt{\rho^2-\left(\frac{t}{2}\right)^2}=\sqrt{RR_0} \tag{6.44}$$

The importance of this strain reversal is that it must be taken into account in determining the effective strain in a material element at a particular distance from the middle surface. If there is no reversal, the effective strain can be determined approximately from the initial and final lengths of the fibre. If there is a reversal, the effective strain integral should be integrated along the whole deformation path. In Figure 6.21(c), the bold curve on the right shows the effective strain determined from the above analysis. The broken line is that determined from the simple, large bend ratio analysis. There is a significant difference due to two factors:

- appropriate integration of the effective strain; and
- the non-linear distribution of strain derived from Equation 6.41.

The curves in Figure 6.21(c) are calculated for a bend ratio $\rho/t = 2/3$. At the inner surface, $y_0 = -t/2$, the strain calculated from Equation 6.41 is $\varepsilon_1 = -1.4$. In the simple analysis

$$\varepsilon_1=\frac{y}{\rho}=\frac{-t}{2\rho}=-0.75$$

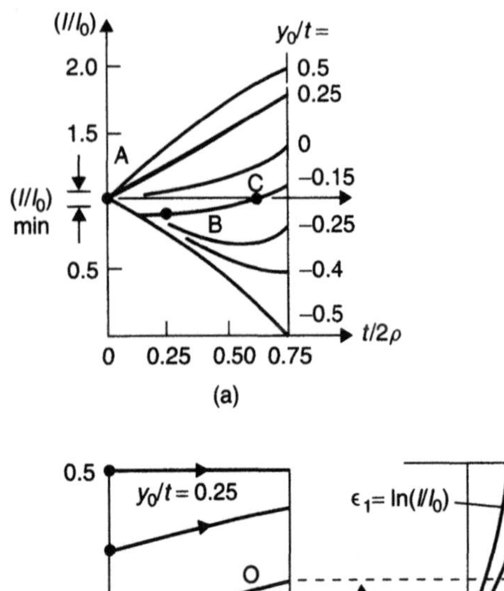

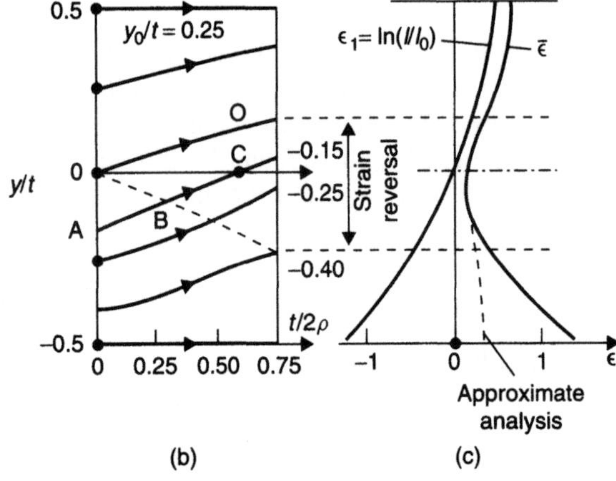

Figure 6.21 (a) Changes in length of fibres at different initial distances y_0/t from the middle surface. (b) Changes in distance from the middle surface during bending. (c) Axial strain ε_1 and effective strain $\bar{\varepsilon}$ in sheet bent to a bend ratio of $\rho/t = 2/3$; the effective strain derived from the simple, large bend ratio analysis is shown by the broken line.

6.7.2 Stress distribution in small radius bends

In the simple analysis, the stresses assumed in the bend region are those given by Equation 6.5. In small radius bends, the through-thickness stress cannot be assumed to be zero. As plane strain is assumed, the strain and stress state, as shown in Figure 6.22, is

$$\varepsilon_1; \quad \varepsilon_2 = 0; \quad \varepsilon_3 = -\varepsilon_1$$

$$\sigma_1; \quad \sigma_2 = (\sigma_1 + \sigma_3)\,/\,2; \quad \sigma_3 \tag{6.45}$$

The equilibrium equation for forces in the through-thickness (radial) direction is

$$(\sigma_3 + \mathrm{d}\sigma_3)\,(r + \mathrm{d}r)\,\mathrm{d}\theta 1 - (\sigma_3 r\;\mathrm{d}\theta + \sigma_1\;\mathrm{d}r\;\mathrm{d}\theta)\;1 = 0$$

i.e.

$$\frac{\mathrm{d}\sigma_3}{\mathrm{d}r} - \frac{\sigma_1 - \sigma_3}{r} = 0 \tag{6.46}$$

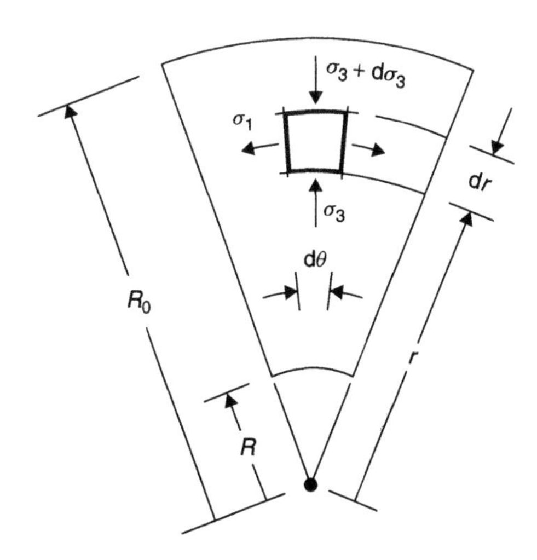

Figure 6.22 Stresses on an element at a radius r in a small radius bend.

To investigate the stresses, we assume a Tresca yield criterion as in Section 2.4.4. In the region where the bending stress is tensile, the greatest stress is σ_1 and the least is σ_3; therefore,

$$(\sigma_1 - \sigma_3) = S = \sigma_f$$

and Equation 6.46 becomes

$$\frac{d\sigma_3}{dr} - \frac{S}{r} = 0$$

If the stress on the outer surface is zero, integrating between the limits $r = R_0$ and r, we obtain

$$\sigma_3 = -S \ln \frac{R_0}{r} \tag{6.47a}$$

In the region where the bending stress is compressive, the yield criterion gives that

$$(\sigma_1 - \sigma_3) = -S = -\sigma_f$$

Integrating Equation 6.46 between the limits $r = R$ and r, and assuming that the stress on the inner surface is zero, we obtain

$$\sigma_3 = -S \ln \frac{r}{R} \tag{6.47b}$$

The stress distributions are shown schematically in Figure 6.23. It should be noted that these are for a non-strain-hardening material in which S is constant. The through thickness stresses given by Equations 6.67(a) and (b) are equal at a radius of

$$r_b = \sqrt{RR_o}$$

As noted earlier, this is the radius at which the bending strain reverses.

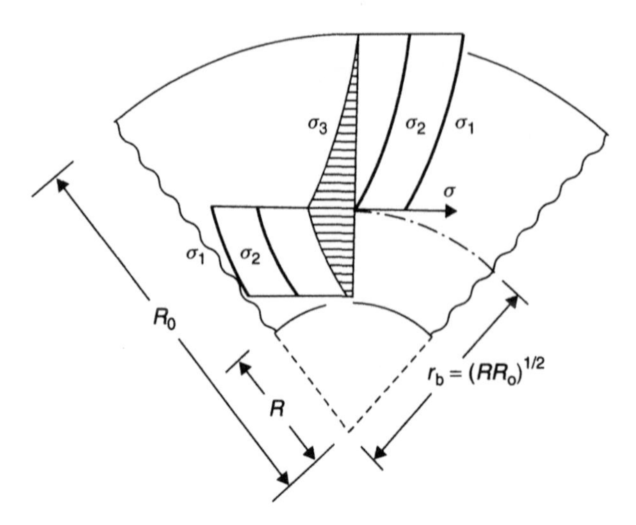

Figure 6.23 Stress distributions in bending a rigid, perfectly plastic sheet to a small bend ratio.

In determining the stress distributions for a strain-hardening material it is necessary to determine the strain ratios in each element and from these evaluate the effective strain as outlined in the preceding section. The result for a material obeying a strain-hardening stress strain law and bent to a bend ratio of 2.5 ($\rho = 2.5t$) is shown in Figure 6.24.

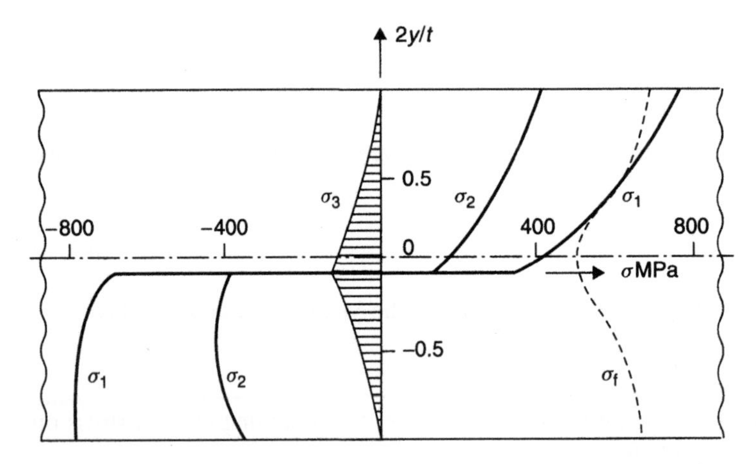

Figure 6.24 Stress distributions for a rigid, strain-hardening sheet bent to a bend ratio of 2.5.

6.8 The bending line

6.8.1 The moment curvature characteristic

The determination of the shape of a sheet that is bent under the action of a force or moments depends on knowing the moment curvature characteristic. Examples of this are shown in the preceding section, for example in Figure 6.17. This shows a characteristic determined by analysis, but techniques also exist to obtain such a diagram experimentally. In studying practical problems it is highly desirable to use a moment diagram determined experimentally as inaccuracies may exist in curves calculated from tensile test data due, among other things, to:

- inaccuracies in data from tensile test at small strains in the region of the elastic/ plastic transition;
- the yield criterion adopted being only an approximation;
- variation of properties through the thickness of the sheet; and
- anisotropy in the sheet that is not well characterized.

Assuming that a reasonable moment curvature characteristic is available, the determination of the bent shape of the sheet is often time-consuming. In this section, a construction is described called *the bending line* that will give quickly some of the information needed.

6.8.2 The bending line construction

In Figure 6.25, a sheet bent by a horizontal force per unit width of sheet P is shown on the left. On the right, a scaled version of the moment curvature characteristic is drawn such that the curvature axis is collinear with the line of action of the force. The ordinate of the curve is obtained from the moment curvature diagram by dividing the moment by the force P. At some point a distance s along the sheet at a height a the moment is $M = Pa$, and hence

$$a = \frac{M}{P} \tag{6.48}$$

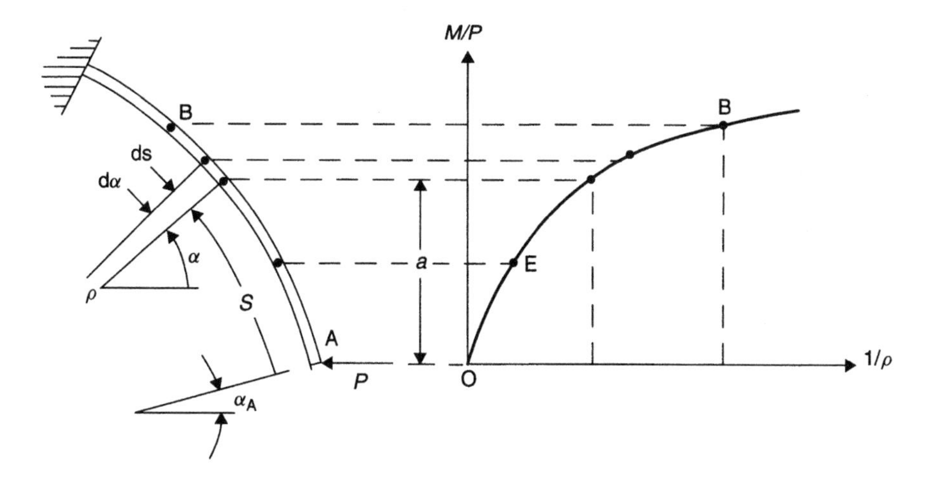

Figure 6.25 Diagram of the deflected shape of a sheet bent by a line force P per unit length on the left, and the scaled moment curvature characteristic on the right.

Thus, a is both the height of the point above the line of application of the force A and the appropriate distance along the scaled moment curvature diagram. (It should be noted that as the units of M are [force][length]/[length] and of P [force]/[length], the unit of M/P is [length].)

This construction permits the curvature at any point to be obtained directly. For example, at point B on the curved sheet, the curvature is given by the corresponding point B in the diagram on the right.

The angle of the normal to the bent sheet, α, can also be found from this construction. As shown on the left-hand side of Figure 6.25, the change in the angle of the normal over a small distance along the sheet ds is dα. The corresponding change in height is da, where

$$\mathrm{d}a = \mathrm{d}s \cos\alpha$$

As $\mathrm{d}s = \rho \mathrm{d}\alpha$, we obtain

$$\cos\alpha \, \mathrm{d}\alpha = \left(\frac{1}{\rho}\right) \mathrm{d}a \tag{6.49}$$

Integrating between the points A and B on the sheet, we obtain

$$\int_{\alpha_{\mathrm{A}}}^{\alpha_{\mathrm{B}}} \cos\alpha \, \mathrm{d}\alpha = \sin\alpha_{\mathrm{B}} - \sin\alpha_{\mathrm{A}} = \int_{\mathrm{o}}^{a_{\mathrm{B}}} \frac{1}{\rho} \, \mathrm{d}a \tag{6.50}$$

The term on the right is the area between the curve OEB and the vertical axis. This can be found by a graphical method from an experimental moment curvature diagram or calculated from a theoretically determined characteristic. Thus, knowing the direction of the normal at A, the direction of the normal at B can be determined.

6.8.3 Examples of deflected shapes

It is often useful to know the region in a bent sheet where the deformation will be elastic. If the elastic/plastic transition is known in the moment curvature diagram, e.g. at point E on the curve in Figure 6.25, the sheet will be elastically deformed between A and the height corresponding to E. On release of the force, this portion of sheet will spring back to a straight line.

The effect of material properties on the deformed shape of sheets bent by line forces is illustrated in Figure 6.26. In (a), the material is rigid, perfectly plastic. The area between the moment curve and the axis C between A and B is zero and therefore, from Equation 6.50, the normals to the sheet at these points are parallel and the sheet is straight. There is a *plastic hinge* at B. The value of the force P is uniquely determined as

$$\frac{M_{\mathrm{p}}}{P} = a_{\mathrm{B}}$$

In Figure 6.26(b) the moment curve for a strain-hardening sheet is shown on the right. The differences between the sines of the angles of the normals is given by the area between the curve OB and the axis OC. In Figure 6.26(c), a linear moment curve is shown and the

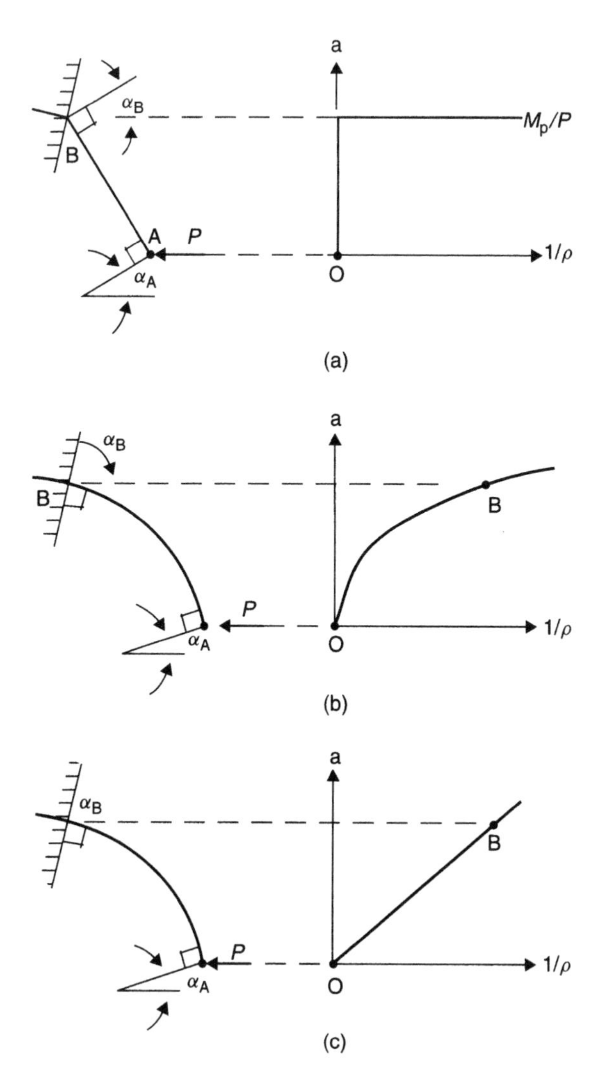

Figure 6.26 Bending line construction for a sheet bent by a horizontal line force for (a) a rigid, perfectly plastic sheet, (b) a strain-hardening sheet, and (c) a sheet having a linear stress–strain relation.

curvature of the sheet will be proportional to the height above the line of application of the force.

In Figure 6.27, a sheet is bent between two smooth, parallel plates. This is similar to the operation of *hemming* where the edge of a sheet is bent over itself or another sheet. If we assume that the force is applied at the tangent point and is normal to the plates, then the construction is shown in the diagram. The curvature will vary from zero at the tangent point to a maximum at the nose.

In the manufacture of articles such as hose clips, a strip may be bent over a form roll as shown in Figure 6.28. The curvature of the strip should match that of the form roll at the tangent point, i.e. $(1/\rho_B) = 1/R$. If the force to bend the strip is applied by a small roller at A and at the instant shown the strip is horizontal, then the angle of the normal at

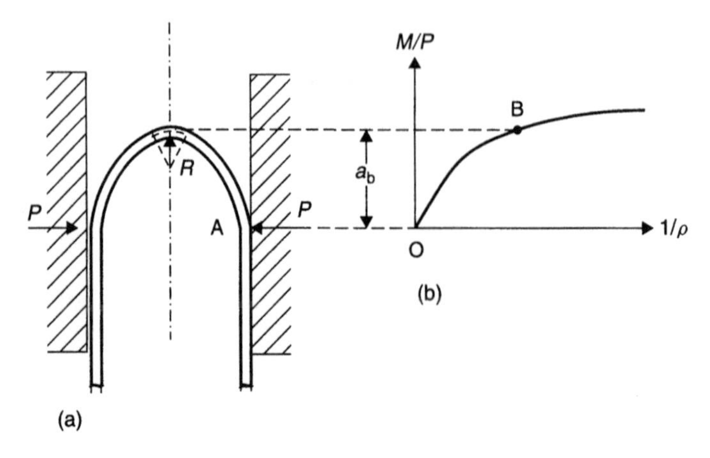

Figure 6.27 Bending a strip between two parallel, frictionless dies.

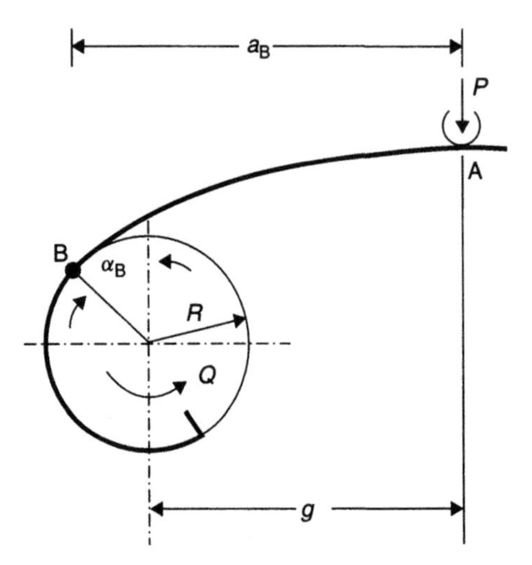

Figure 6.28 Process for curving a strip over a form roll of radius R.

A is zero and at B is α_B. For a given moment curvature characteristic, the distance, g, at which the roller should be located to satisfy the above conditions can be found.

6.9 Bending a sheet in a vee-die

A common process for bending a sheet in a press is shown in the Introduction in Figure I. 3(b) and in more detail in Figure 6.29. If the angle of the die face is α in Figure 6.29(a) and there is some friction between the sheet and the die, the force on the sheet will be at an angle ψ to the normal, where the coefficient of friction is $\mu = \tan\psi$. The force on the punch is

$$P = 2\cos(\alpha - \psi) \qquad (6.51)$$

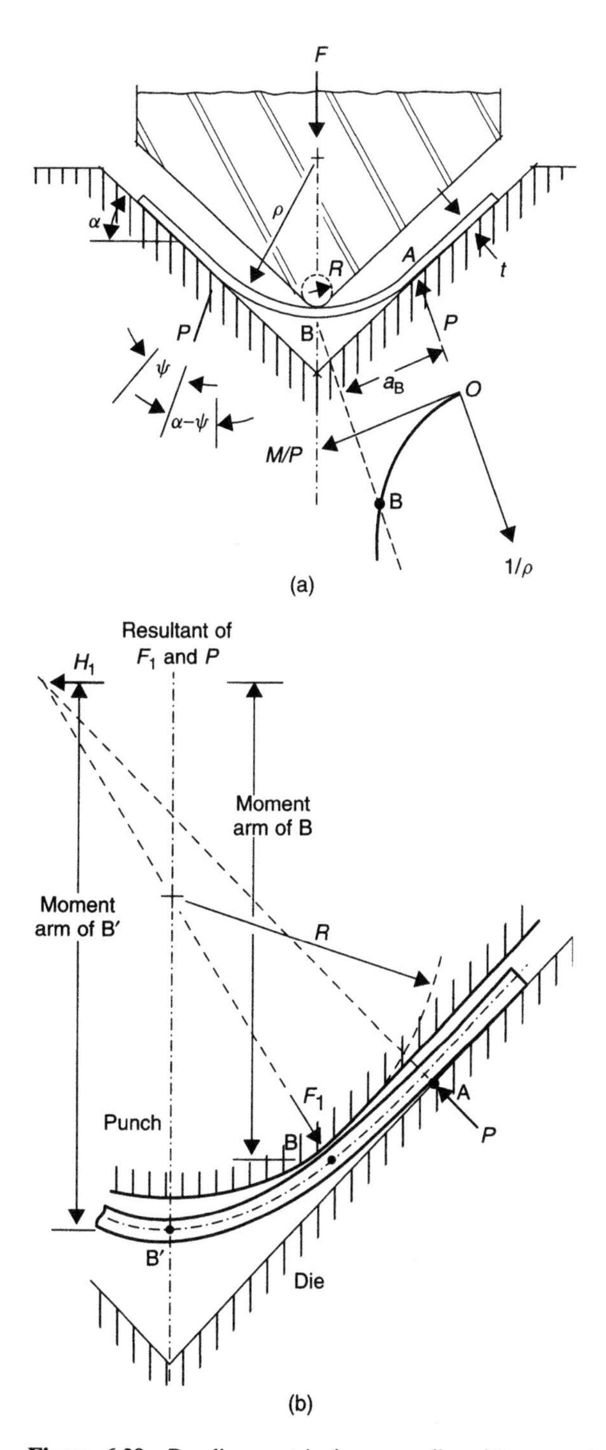

Figure 6.29 Bending a strip in a vee-die with a punch of nose radius R. (a) At the start of the process. (b) When the punch has nearly reached the bottom of its stroke.

In the initial stages, the curvature of the sheet at the nose of the punch will be less than the nose radius, as shown in Figure 6.29(a). The curvature is given by the point B in the bending line construction shown. As the bending progresses, the punch force will increase and the curvature at the point of contact increase until it just matches the punch curvature. On further bending, the point of contact with the punch will move away from the nose to some point B as shown in Figure 6.29 (b). Here we consider only a frictionless condition, so that the force is normal to the tool at the point of contact. It is seen that there is a difference between the line of action of the force exerted by the die at point A and that through the point of contact B with the punch. These forces converge as shown, and by symmetry, their resultant must be horizontal, i.e. the force H. As the moment arm of the force bending the sheet at the centre-line B′ is greater than that at the punch contact B, the curvature at B′ must be greater than at B, and there will be a gap between the sheet and the punch at B′.

If close conformity between the punch and the sheet is required, the vee-die is made with a radius at the bottom to match the punch and a large force is applied at the end of the process. A problem with such an arrangement is that small variations in thickness or strength in the sheet or in friction may cause appreciable changes in springback.

6.10 Exercises

Ex. 6.1 A steel strip, 50 mm wide and 2 mm thick is bent by a pure moment. The plane strain–stress curve follows the law $\sigma = 200\varepsilon$ GPa in the elastic range, and $\sigma = 250$ MPa (constant) in the plastic range.

(a) Determine the limiting elastic curvature.

(b) Construct a moment curvature diagram for the strip in the range $0 < \left(\frac{1}{\rho}\right) < 8\,\text{m}^{-1}$

(c) If the strip is wound without tension onto a former of 800 mm diameter, determine the radius of curvature after release:

- using the approximate relation $\Delta\left(\frac{1}{\rho}\right) = -3\frac{\sigma_f}{Et}$;
- using the diagram in (b).

[*Ans* : (*a*)*1.25m*$^{-1}$, (*c*)*1.6 m vs.1.28 m.*]

Ex. 6.2. A strip of length l and thickness t is bent to a complete circle. Under load the circumference is l. The material is elastic, perfectly plastic with a constant plane strain yield stress S and elastic modulus E'. Calculate the gap between the ends of the strip after unloading. Assume that the strip is fully plastic when bent and that the final gap is small.

$$\left[Ans : \frac{3}{2\pi}\frac{l^2}{t}\frac{S}{E'}\right]$$

Ex. 6.3 A strip of sheet steel, 2 mm thick and 200 mm wide, is to be bent in a die under conditions of zero friction and zero axial tension. The radius of curvature of the die is 80 mm. The plastic properties of the material are $\overline{\sigma} = 600\overline{\varepsilon}^{0.22}$ MPa. The elastic properties are $E = 200$ GPa and Poissons ratio = 0.3.

(a) Find the radius after springback.

[*Ans* : *92 mm*]

Ex. 6.4 Springback in bending is given by Equation 6.32. If Aluminum is substituted for steel in identical applications with the same bend radius and sheet thickness, compare the difference in springback for bend ratio of (a) 10 and (b) 5 respectively.

Al: $K = 205\,\text{MPa}$, $n = 0.2$, density: $\rho = 2.7 \times 10^3\,\text{kg/m}^3$; $E = 75\,\text{GPa}$, $\nu = 0.3$.

Steel: $K = 530\,\text{MPa}$, $n = 0.26$, density: $\rho = 7.9 \times 10^3\,\text{kg/m}^3$; $E = 190\,\text{GPa}$, $\nu = 0.3$.

[*Ans : Springback ratio, Al to Steel:(a)1.173, (b)1.125.*]

7
Simplified analysis of circular shells

7.1 Introduction

Many sheet forming operations can be modelled as the deformation of circular (axisymmetric) shells. Examples are stretching a sheet over a domed punch, drawing a circular disc to form a cylindrical cup and tube forming processes such as flaring and necking. Detailed analyses usually require numerical methods, but by making certain simplifying assumptions, useful information about forming loads and strains can obtained from approximate analytical models. This chapter introduces the basic simplifying assumptions and some applications to typical forming processes.

7.2 The shell element

An element of an axisymmetric shell is shown in Figure 7.1. The surface element is at a radius r and subtends an angle $d\theta$. The thickness is t and the principal stresses are σ_θ in the hoop direction and σ_ϕ along the meridian; the radial stress perpendicular to the element is considered small so that the element is assumed to deform in plane stress. Also acting on the element are the principal tensions, $T_\theta = \sigma_\theta t$ and $T_\phi = \sigma_\phi t$. If the element is deforming plastically, the principal stresses will satisfy the yield condition and here we select the Tresca criterion. It follows that in a region in which the thickness is uniform, the tensions will also satisfy a similar condition, and this is illustrated for plane stress, in Figure 7.2. This may be considered as a tension yield locus and following an approach similar to that in Section 3.7, we identify an *effective* or *representative tension* function $\overline{T}$.

The major simplifying assumption employed here is that the yielding tension $\overline{T}$ in Figure 7.2 will remain constant throughout the process. This implies that strain-hardening will balance material thinning, i.e. $\sigma_f t = \overline{T}$, is constant. This happens to be a better assumption than neglecting strain-hardening. The reason can be seen by reference to Figures 3.3(a) and 5.17. In drawing processes (along the left-hand diagonal) the material does not change thickness and it is preferable to use a non-strain-hardening sheet as there is no danger of necking; strain-hardening would only increase the forming loads and make the process more difficult to perform. On the other hand, in stretching processes that lie in the

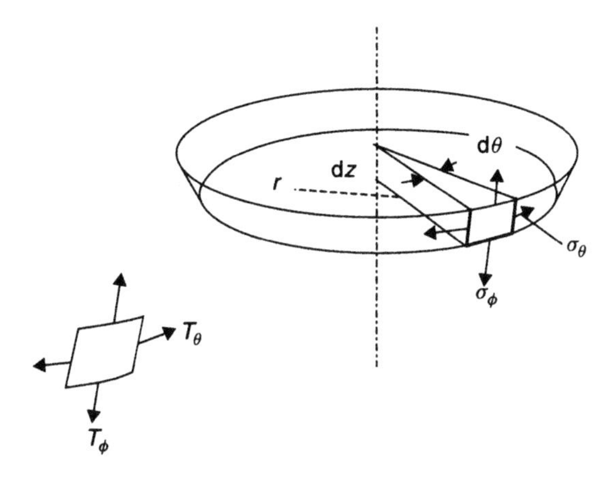

Figure 7.1 An element of an axisymmetric shell.

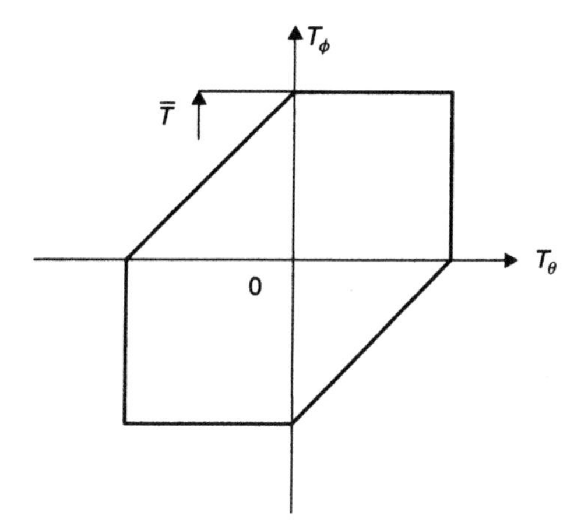

Figure 7.2 Yield diagram for principal tensions where the locus remains of constant size and the effective tension $\overline{T}$ is constant.

first quadrant, strain-hardening is needed in the sheet to avoid local necking and tearing. As discussed in Section 3.3.1, thinning will accompany stretching processes and while the stresses increase due to strain-hardening, the sheet will thin rapidly and, to a first approximation, the product of stress and thickness will be constant.

7.2.1 Shell geometry

An axisymmetric shell, or surface of revolution, is illustrated in Figure 7.3(a). A point on the surface, P, can be described in terms of the cylindrical coordinates r, θ, z as shown. The curve generating the shell, C, is illustrated in Figure 7.3(b) and the outward normal to the curve (and the surface) at P is $N\vec{P}$. This makes an angle ϕ with the axis. The ordinary curvature of the curve at P is ρ_2, and this is also one of the principal radii of curvature of the surface. The other principal radius of curvature of the surface is ρ_1, as shown. The arc

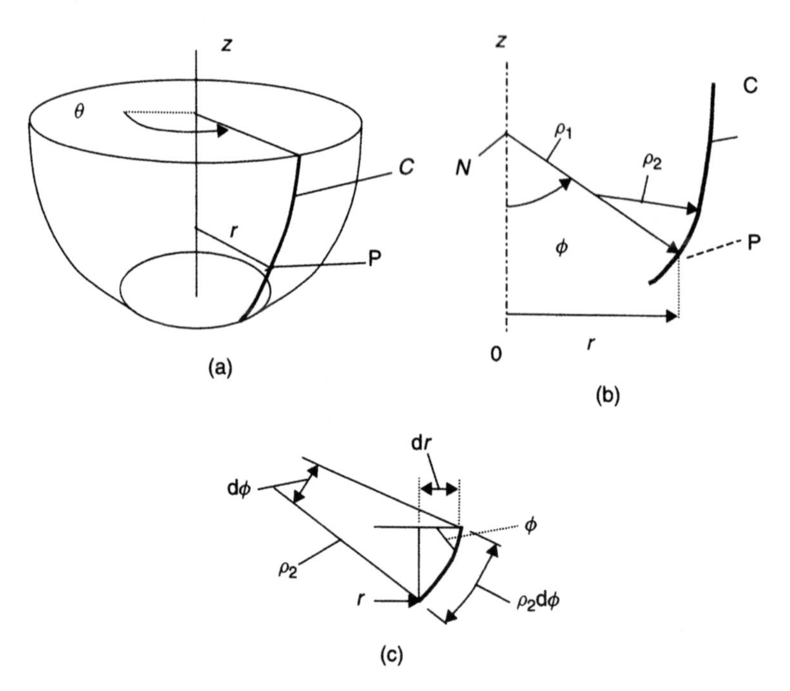

Figure 7.3 (a) Surface of revolution swept out by rotation of a curve C about the z axis. (b) Principal radii of curvature at the point P. (c) Geometric relations at P.

length of the element along the meridian is $ds = \rho_2\, d\phi$, and from Figure 7.3(b) and(c), the following geometric relations can be identified,

$$r = \rho_1 \sin\phi \tag{7.1}$$

and

$$dr = \rho_2\, d\phi \cos\phi \tag{7.2}$$

7.3 Equilibrium equations

As shown in Figure 7.4, we consider a shell element of sides $r\, d\theta$ and, $\rho_2\, d\phi$. The pressure acting on this element exerts an outward force along the surface normal of

$$pr\, d\theta \rho_2\, d\phi$$

Due to the curvature of the shell, the forces on the element $T_\theta \rho_2\, d\phi$ in the hoop or circumferential direction exert an inward force in the horizontal direction of

$$T_\theta \rho_2\, d\phi\, d\theta$$

The component of this force along the normal is

$$T_\theta \rho_2\, d\phi\, d\theta \sin\phi$$

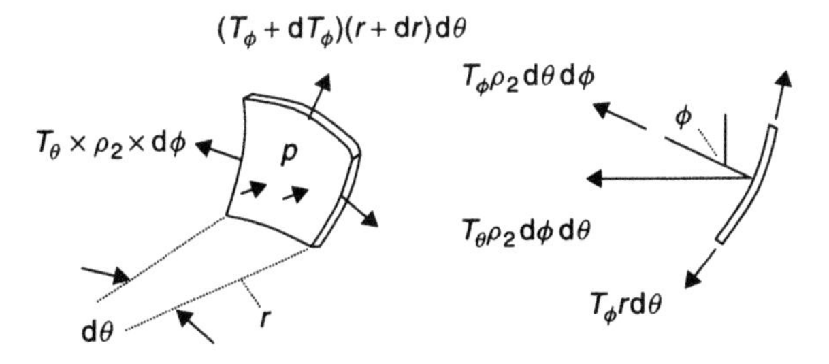

Figure 7.4 Forces acting on a shell element.

and the component tangential to the surface in the direction of the meridian is

$$T_\theta \rho_2 \, d\phi \, d\theta \cos\phi$$

Due to the curvature of the shell the forces along the meridian $T_\phi r \, d\theta$ exert a force in the direction normal to the surface of

$$T_\phi r \, d\theta \, d\phi$$

The equilibrium equation in the direction normal to the surface is

$$pr \, d\theta \rho_2 \, d\phi = (T_\theta \rho_2 \, d\phi \, d\theta) \sin\phi + T_\phi r \, d\theta \, d\phi$$

Combining with Equation 7.1, this reduces to

$$p = \frac{T_\theta}{\rho_1} + \frac{T_\phi}{\rho_2} \tag{7.3}$$

The equilibrium equation in the direction of the meridian is

$$(T_\phi + dT_\phi)(r + dr)d\theta - T_\phi r \, d\theta - T_\theta \rho_2 \, d\phi \, d\theta \cos\phi = 0$$

Combining with Equation 7.2, this reduces to

$$\frac{dT_\phi}{dr} - \frac{T_\theta - T_\phi}{r} = 0 \tag{7.4}$$

7.4 Approximate models of forming axisymmetric shells

Analytical models of some sheet forming processes are developed here using a number of simplifying assumptions. These are summarized as below.

- The shell is symmetric about the central axis and all variables such as thickness, stress and tension are constant around a circumference.
- The thickness is small and all shear and bending effects are neglected.
- Contact pressure between the tooling and the sheet is small and friction is negligible.
- The shell is bounded by planes normal to the axis and boundary loads are uniform around the circumference and act tangentially to the surface of the shell. There are no shear forces or bending moments acting on the boundaries. The total force acting at a

boundary will therefore be a central force along the axis, as shown in Figure 7.5. The magnitude of the axial force is,

$$Z = (T_\phi)_0 2\pi r_0 \sin\phi_0 \tag{7.5}$$

where the subscript zero refers to the boundary conditions.

- At any instant during forming, all elements of the shell are assumed to be deforming plastically.
- The sheet obeys a plane stress Tresca yield condition and, as mentioned, strain-hardens in such a way that the product of flow stress and thickness remains constant, i.e. $\sigma_f t = \overline{T} =$ constant.

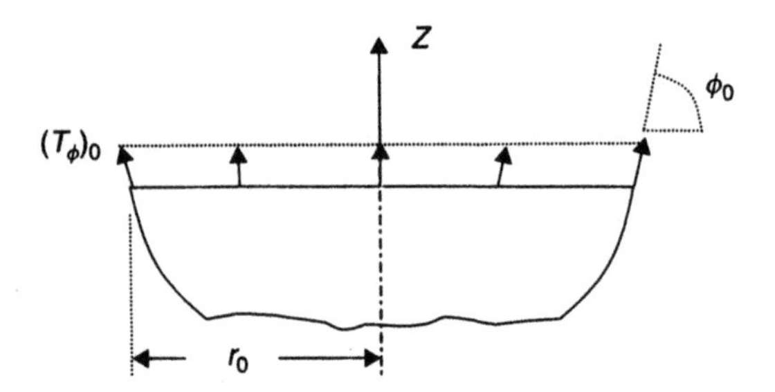

Figure 7.5 Boundary force conditions for a shell.

7.5 Applications of the simple theory

7.5.1 Hole expansion

We consider a circular blank stretched over a domed punch as shown in Figure 7.6(a). At any instant, there is a circular hole at the centre of radius r_i, and a meridional tension is applied at the outer radius r_0. At the edge of the hole, the meridional tension must be zero and a state of uniaxial tension in the circumferential direction would exist. We expect that the meridional tension would become more tensile towards the outer edge and in the yield locus, the tensions would fall in the first quadrant of the diagram as illustrated in Figure 7.6(b).

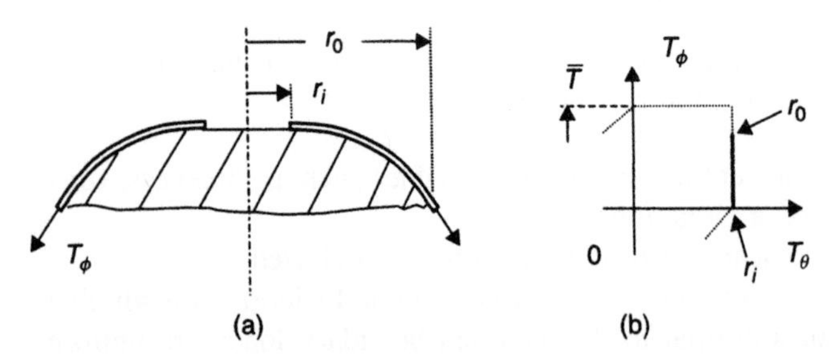

Figure 7.6 (a) Hole expansion process with the sheet stretched over an axisymmetric punch. (b) Region on the tension yield locus for this process.

As seen from Figure 7.6(b), the tensions in the sheet are T_ϕ and $T_\theta = \overline{T}$. The equilibrium equation, Equation 7.4, is

$$\frac{dT_\phi}{dr} - \frac{\overline{T} - T_\phi}{r} = 0 \tag{7.6}$$

which on integrating and substitution of the boundary condition, $T_\phi = 0$ at $r = r_i$, gives

$$T_\phi = \overline{T}\left(1 - \frac{r_i}{r}\right) \tag{7.7}$$

The tension distribution given by Equation 7.7 is illustrated in Figure 7.7.

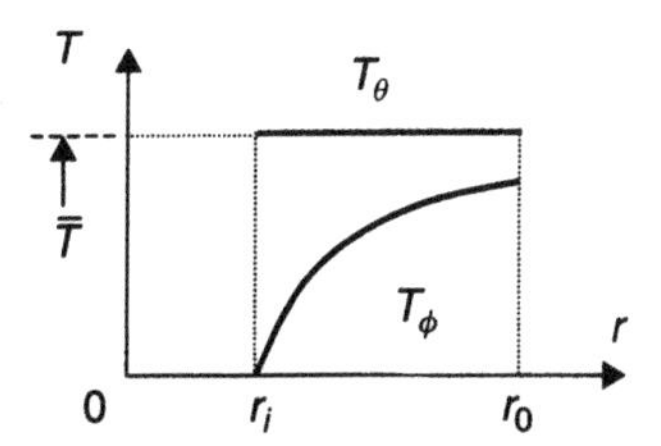

Figure 7.7 Stress distribution for hole expansion in a circular blank.

For a circular blank without a central hole, the stress state at the pole is, by symmetry, that $T_\phi = T_\theta$. From Figure 7.6(b) it is seen that this can only occur when both are equal to the yield tension $\overline{T}$. The equilibrium equation is then

$$\frac{dT_\phi}{dr} = 0$$

i.e. the merdional tension does not change with radius and hence the stress distribution is uniform and

$$T_\phi = T_\theta = \overline{T}$$

This relation is useful for determining the punch load. If, instead of being stretched over a punch, the sheet is clamped around the edge and bulged by hydrostatic pressure, the hoop strain around the edge will be zero and the hoop tension at the outer edge will be less than the meridional stress. In this case, the simple model does not predict the tension distribution well; at the edge, the strain state must be plane strain and from the flow rule, Equation 2.13(c), we predict that $T_\theta = T_\phi/2$.

7.5.2 Drawing

If a circular blank is drawn into a circular die as shown in Figure 7.8(a), we may anticipate that the meridional tension will be tensile (positive) at the throat and zero at the outer edge. As any circumferential line will shrink during drawing, the hoop tensions are likely to be negative or compressive. The tensions will therefore lie in the second quadrant of the yield locus as shown in Figure 7.8(b) where

$$T_\phi - T_\theta = \overline{T} \tag{7.8}$$

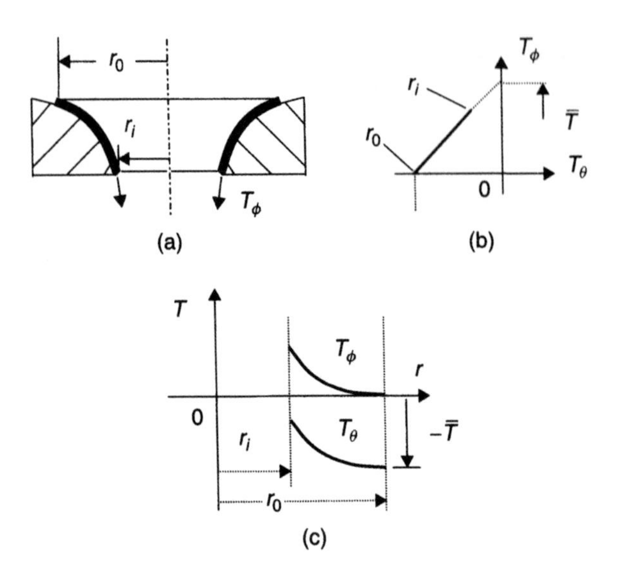

Figure 7.8 (a) Drawing of a circular shell. (b) Location of tensions on the yield locus. (c) Distribution of tensions in the shell.

The equilibrium equation is then

$$\frac{\mathrm{d}T_\phi}{\mathrm{d}r} + \frac{\overline{T}}{r} = 0 \tag{7.9}$$

Integrating and substituting the boundary condition that $T_\phi = 0$ at $r = r_0$, we obtain,

$$T_\phi = \overline{T}\ln\frac{r_0}{r} \quad \text{and} \quad T_\theta = T_\phi - \overline{T} = -\overline{T}\left(1 - \ln\frac{r_0}{r}\right) \tag{7.10}$$

This stress distribution is illustrated in Figure 7.8(c).

It may be seen from Figure 7.8(b) that the maximum value for the meridional tension at the inner radius of the drawn shell is when $T_\phi = \overline{T}$. Substituting this in Equation 7.10 gives that the maximum size shell that can be drawn is when

$$\ln\frac{r_0}{r_\mathrm{i}} = 1 \quad \text{or} \quad \frac{r_0}{r_\mathrm{i}} = \mathrm{e} = 2.72 = LDR \tag{7.11}$$

This so-called *Limiting Drawing Ratio* (LDR) given by the simple analysis is very approximate and actual values in the range of 2.0–2.2 are usually observed.

7.5.3 Nosing and flaring of tube

The end of a tube may be 'nosed' or 'necked' by pushing into a converging die as shown in Figure 7.9(a) or 'flared' using a cone punch as shown in Figure 7.9(b). In nosing, the tensions lie in the third quadrant of the yield locus, Figure 7.9(c), and it may be shown that, for the boundary condition $T_\phi = 0$ at r_i the tensions are

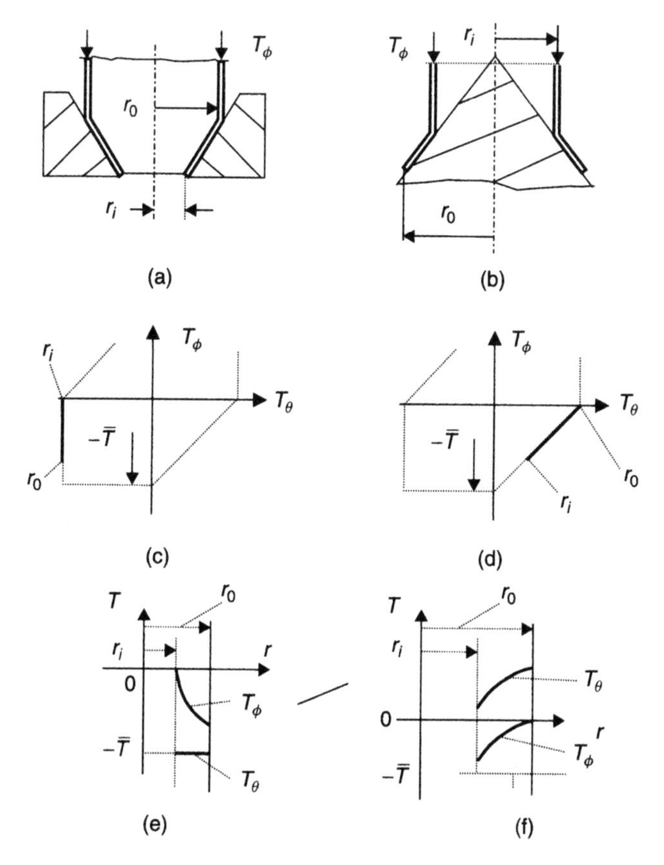

Figure 7.9 (a) Nosing and (b) flaring the end of a tube. Location of the tensions on the yield locus for (c) nosing and (d) flaring. Distribution of tensions for (e) nosing and (f) flaring.

$$T_\phi = -\overline{T}\left(1 - \frac{r_i}{r}\right) \quad \text{and} \quad T_\theta = -\overline{T} \tag{7.12}$$

These tensions are illustrated in Figure 7.9(e).

In flaring the tensions lie in the fourth quarter as shown in Figure 7.9(d) and, for the boundary condition $T_\phi = 0$ at r_0 the tensions are

$$T_\phi = -\overline{T}\ln\frac{r_0}{r} \quad \text{and} \quad T_\theta = \overline{T} + T_\phi = \overline{T}\left(1 - \ln\frac{r_0}{r}\right) \tag{7.13}$$

These tensions are illustrated in Figure 7.9(f).

7.6 Summary

The equilibrium equation for forces in the direction normal to the shell surface, i.e.

$$p = \frac{T_\theta}{\rho_1} + \frac{T_\phi}{\rho_2} \tag{7.3}$$

is always valid, but the equation for forces tangential to the sheet,

$$\frac{dT_\phi}{dr} - \frac{T_\theta - T_\phi}{r} = 0 \tag{7.4}$$

neglects the effect of friction between the sheet and the tooling. In integrating this equation, we have assumed that the effective tension $\overline{T}$ is constant. This introduces further error, although as mentioned, it is usually a better assumption than assuming that the material is rigid, perfectly plastic.

If the material obeys a stress strain law $\overline{\sigma} = K(\varepsilon_0 + \overline{\varepsilon})^n$, an appropriate value for $\overline{T}$ could be

$$\overline{T} = K\varepsilon_0^n t_0 \tag{7.14}$$

where t_0 is the initial thickness. If ε_0 is very small, Equation 7.14 would probably underestimate the effective tension and some discretion should be exercised.

7.7 Exercises

Ex. 7.1 In a hole expansion process, the inner edge is unloaded and the meridional tension at the outer radius ($r = r_0$) is $T_\phi = 2\overline{T}/3$. If the sheet is fully plastic, what is the current ratio (r_0/r)?
[*Ans : 3*]

Ex. 7.2 For the nosing operation shown in Figure 7.9 (a), given the boundary condition $T_\phi = 0$ at r_i, show that the meridional tension is distributed as follows:

$T_\phi = -\overline{T}\left(1 - \frac{r_i}{r}\right)$ and $T_\theta = -\overline{T}$.

Ex. 7.3 In a flaring operation, what is the range of r for which the equation

$T_\phi = \overline{T}\ln\left(\frac{r_0}{r}\right)$ is valid?

[*Ans* : $r_i \le r \le er_i$]

8

Cylindrical deep drawing

8.1 Introduction

In Chapter 7, a simple approach to the analysis of circular shells was given. Here we examine in greater detail the deep drawing of circular cups as shown in Figure 8.1. This can be viewed as two processes; one is stretching sheet over a circular punch, and the other is drawing an annulus inwards. The two operations are connected at the cylindrical cup wall, which is not deforming, but transmits the force between both regions.

The simple analysis in Section 7.5.2 gave a limit to the size of disc that could be drawn as e(= 2.72) times the punch diameter. This over-estimates the Limiting Drawing Ratio and in this chapter we investigate various factors that influence the maximum blank size.

In the single-stage process shown in Figure 8.1, the greatest ratio of height to diameter in a cup is usually less than unity; this is determined by the Limiting Drawing Ratio. Deeper cups may be made by redrawing or by thinning the cup wall by ironing and these processes are studied.

8.2 Drawing the flange

The flange of the shell can be considered as an annulus as shown in Figure 8.2; the stresses on an element at radius r are shown in Figure 8.3. The equilibrium equation for this element is, in the absence of friction,

$$(\sigma_r + \mathrm{d}\sigma_r)\,(t + \mathrm{d}t)\,(r + \mathrm{d}r)\,\mathrm{d}\theta = \sigma_r t r\,\mathrm{d}\theta + \sigma_\theta t\,\mathrm{d}r\,\mathrm{d}\theta$$

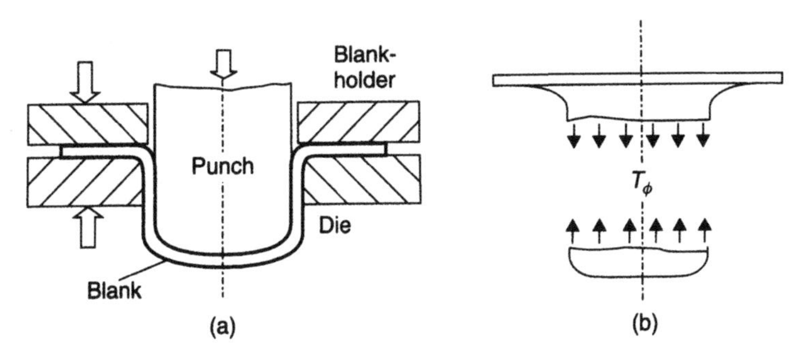

Figure 8.1 (a) Drawing a cylindrical cup from a circular disc. (b) Transmission of the stretching and drawing forces by the tensions in the cup wall.

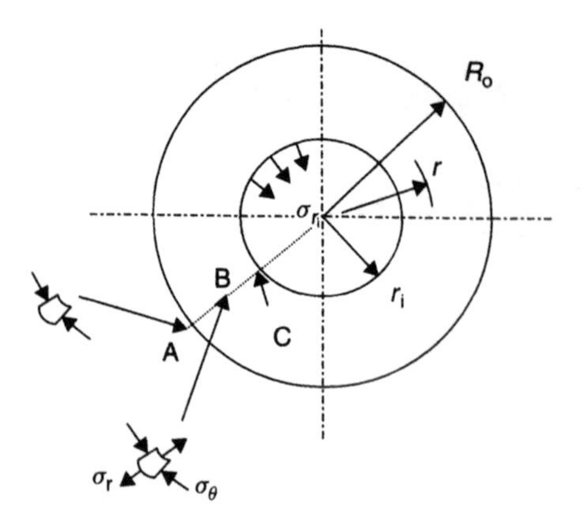

Figure 8.2 Annular flange of a deep-drawn cup.

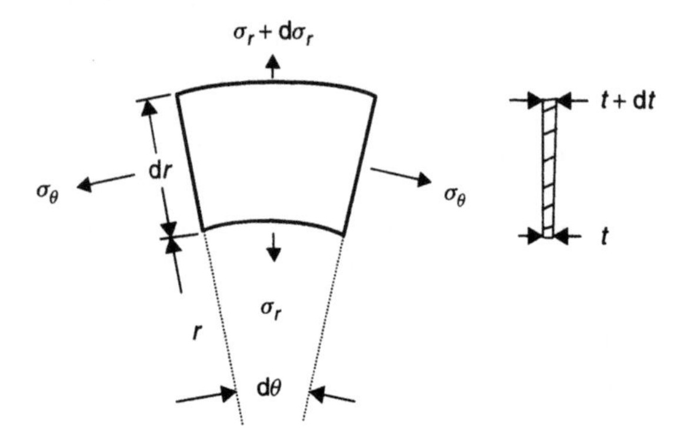

Figure 8.3 Element in the annular flange in Figure 8.2.

which reduces to,

$$\frac{d\sigma_r}{dr} + \frac{\sigma_r}{t}\frac{dt}{dr} - \frac{\sigma_\theta - \sigma_r}{r} = 0 \tag{8.1}$$

At the outer edge, point A, there is a free surface and $\sigma_r = 0$; the stress state is therefore one of uniaxial compression in which $\sigma_\theta = -\sigma_f$, where σ_f is the current flow stress. At some intermediate radius, point B, the radial stress will be equal and opposite to the hoop stress, while at the inner edge, point C, the radial stress will be a maximum. The stress states and the corresponding strain vectors are shown on the von Mises yield locus, Figure 8.4. (This is similar to the diagram, Figure 2.9). At the outer edge A the blank will thicken as it deforms, while at point B there will be no change in thickness. At the inner edge C the sheet will thin. The overall effect is that in drawing, the *total area* of the material initially in the flange will not change greatly. This is a useful approximation when determining blank sizes.

An incremental model can be constructed for drawing the flange using Equation 8.1 and the deformation followed using a numerical method. This is not developed here, but we consider just the initial increment in the process. Using the Tresca yield condition,

$$\sigma_\theta - \sigma_r = -(\sigma_f)_0 \tag{8.2}$$

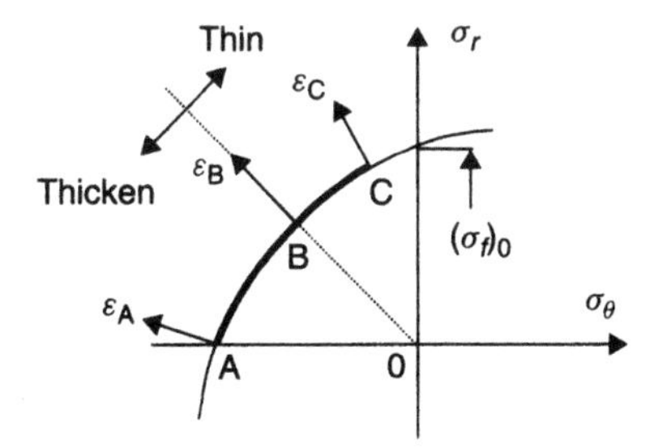

Figure 8.4 Stress state and strain vectors for different points on the flange.

where $(\sigma_f)_0$ is the initial flow stress and as the thickness initially is uniform, i.e. $t = t_0$, Equation 8.1 can be integrated directly. Using the boundary conditions $\sigma_r = 0$ at the outer radius R_o and $\sigma_r = \sigma_{r_i}$, at the inner radius, r_i, we obtain,

$$\sigma_{r_i} = -(\sigma_f)_0 \ln \frac{r_i}{R_o}, \quad \text{and} \quad \sigma_{\theta_i} = -\left\{(\sigma_f)_0 - \sigma_{r_i}\right\} \tag{8.3}$$

For a non-strain-hardening material, the radial stress as given by Equation 8.3 is greatest at the start and will decrease as the outer radius diminishes. The greatest stress that the wall of the cup can sustain for a material obeying the Tresca condition is $(\sigma_f)_0$. Thus substituting $\sigma_{r_i} = (\sigma_f)_0$ in Equation 8.3 indicates that the largest blank that can be drawn, i.e. the *Limiting Drawing Ratio*, has the value

$$\frac{R_o}{r_i} = \mathrm{e} \approx 2.72$$

As indicated, this is an over-estimate, and some reasons for this are mentioned below.

8.2.1 Effect of strain-hardening

Due to strain-hardening, the stress to draw the flange may increase during the process, even though the outer radius is decreasing. As the flange is drawn inwards, the outer radius will decrease and at any instant will be R as shown in Figure 8.5. Due to strain-hardening, the flow stress will increase and become non-uniform across the flange. If we assume an average value $(\sigma_f)_{av.}$ over the whole flange and neglect non-uniformity in thickness, then Equation 8.3 becomes

$$\sigma_{r_i} = (\sigma_f)_{av.} \ln \frac{R}{r_i} \tag{8.4}$$

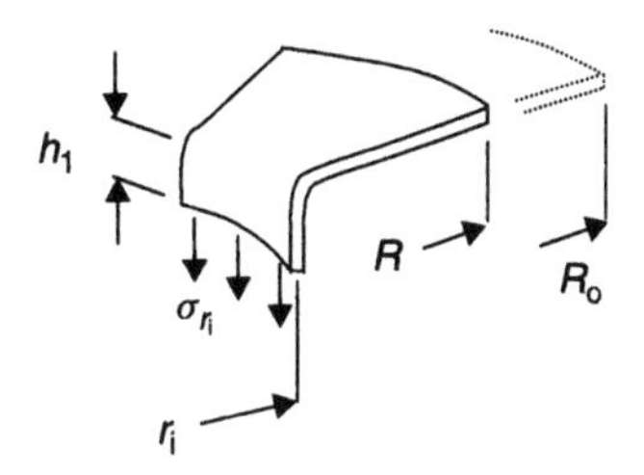

Figure 8.5 Part of a flange during the drawing process for frictionless conditions in which the stress in the wall is equal to the radial stress at the inner radius σ_{r_i}.

There are thus two opposing factors determining the drawing stress, one increasing the stress due to hardening of the material and the other reducing the stress as R becomes smaller. Usually the drawing stress will increase initially, reach a maximum and then fall away as shown in Figure 8.6.

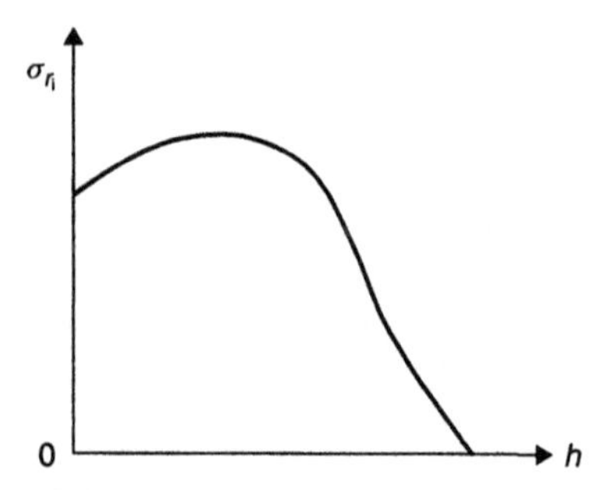

Figure 8.6 Typical characteristic of drawing stress versus punch travel for a strain-hardening material.

8.2.2 Effect of friction on drawing stress

There are two separate ways in which friction will affect the drawing stress. One is at the die radius. Section 4.2.5 gives an analysis of an element sliding around a radius. If the changes in thickness are neglected, Equation 4.13(a) can be written in terms of the stress; i.e.

$$\frac{d\sigma_\phi}{\sigma_\phi} = \mu \, d\phi \tag{8.5}$$

At the die radius, as shown in Figure 8.7, we obtain by integration

$$\sigma_\phi = \sigma_{r_i} \exp \mu \frac{\pi}{2} \tag{8.6}$$

Friction between the blankholder and the flange will also increase the drawing stress. It is a reasonable approximation to assume that the blankholder force B will be distributed around the edge of the flange as a line force of magnitude $B/2\pi R_o$ per unit length, as shown in Figure 8.8. The friction force on the flange, per unit length around the edge, is thus $2\mu B/2\pi R_o$. This can be expressed as a stress acting on the edge of the flange, i.e.

$$(\sigma_r)_{r=R_o} = \frac{\mu B}{\pi R_o t} \tag{8.7}$$

where t is the blank thickness.

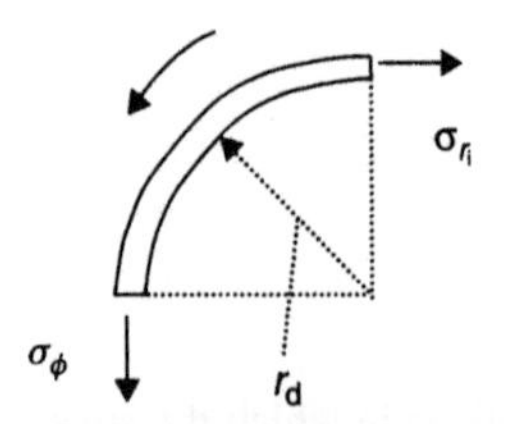

Figure 8.7 Sliding of the flange over the die radius.

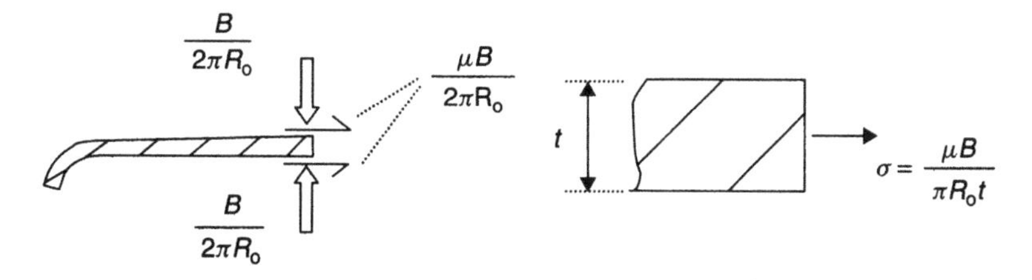

Figure 8.8 Friction arising from the blankholder force, assumed to act at the outer edge.

Both of these factors will increase the stress required to draw the flange. This stress can be determined by various numerical techniques or by approximate models that simplify the effect of strain-hardening and thickness change in the flange

Integrating Equation 8.4 for the new boundary condition, given by Equation 8.7, we obtain

$$\sigma_{r_i} = (\sigma_f)_{av.} \ln \frac{R_o}{r_i} + \frac{\mu B}{\pi R_o t} \tag{8.8}$$

and applying Equation 8.6, the stress in the wall is

$$\sigma_\phi = \frac{1}{\eta} \left\{ (\sigma_f)_{av.} \ln \frac{R}{r_i} + \frac{\mu B}{\pi R t} \right\} \exp \frac{\mu \pi}{2} \tag{8.9}$$

Equation 8.9 is an approximate one that neglects, among other things, the energy dissipated in bending and unbending over the die radius. For this reason, an efficiency factor η has been added; this will have a value less than unity.

8.2.3 The Limiting Drawing Ratio and anisotropy

To determine the Limiting Drawing Ratio some method of determining the average flow stress $(\sigma_f)_{av.}$, the current thickness and the maximum permissible value of wall stress in Equation 8.9 is necessary. This is beyond the scope of this work, but some comments can be made about the influence of different properties on the Limiting Drawing Ratio.

If, in the first instance, we neglect strain-hardening, then the maximum drawing stress will be at the start of drawing; if the flow stress is $\sigma_f = Y = \text{constant}$, the wall stress to initiate drawing is, from Equation 8.9,

$$\sigma_\phi = \frac{1}{\eta} \left\{ Y \ln \frac{R_o}{r_i} + \frac{\mu B}{\pi R_o t_0} \right\} \exp \frac{\mu \pi}{2} \tag{8.10}$$

As indicated, this must not exceed the load-carrying capacity of the wall. If the wall deforms, the punch will prevent circumferential straining, so the wall must deform in plane strain. The stress at which it would deform depends on the yield criterion. Figure 8.9 illustrates yielding in plane strain for various criteria. The line OA indicates the loading path in the cup wall.

For a Tresca yield condition, Figure 8.9(a), the stress in the wall, σ_ϕ, will have the value Y and substitution in Equation 8.10 gives the condition for the maximum blank size, i.e.

$$\left\{ \ln \frac{R_o}{r_i} + \frac{\mu B}{\pi R_o t_0 Y} \right\} \exp \frac{\mu \pi}{2} = \eta$$

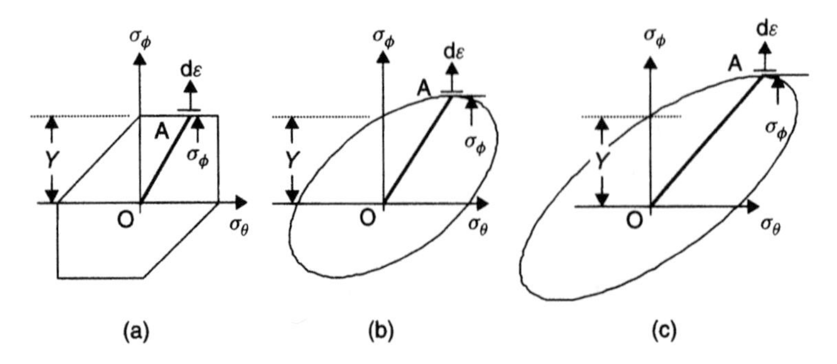

Figure 8.9 Loading path for the cup wall for different yield criteria. (a) The Tresca condition. (b) The von Mises condition. (c) An anisotropic yield locus for a material with an R-value >1.

For the von Mises yield condition (b), the limiting stress in the wall is $\sigma_\phi = (2/\sqrt{3})Y$ and the predicted Limiting Drawing Ratio will be greater. If the material exhibits anisotropy, the yield surface will be distorted. For the case in which the strength of the sheet is higher in the through-thickness direction compared with that in the plane of the sheet, i.e. the R-value is greater than unity, a quadratic yield locus will be elongated along the right-hand diagonal as shown in Figure 8.9(c). The effect is to strengthen the wall, so that a higher stress is required to yield it and therefore the Limiting Drawing ratio will be greater. The increase in LDR predicted from a high exponent yield criterion for a high R value as shown in Section 5.5.5 would be less.

It has been assumed in Figure 8.7 and Equation 8.6 that the thickness of the sheet will not change as it is drawn over the die corner radius. This is not true, and as shown in Section 10.5.2, when sheet is bent or unbent under tension there will be a reduction in thickness. From Equation 10.21, at each bend or unbend, the thickness reduction is

$$\frac{\Delta t}{t} = -\frac{1}{2(\rho/t)}\left(\frac{T}{T_y}\right)$$

where T_y is the tension to yield the sheet and ρ/t the bend ratio at the die corner. Thus a small bend ratio will increase the thickness reduction, reducing the load-carrying capacity of the side-wall and reducing the Limiting Drawing Ratio.

The largest size blank that can be drawn is therefore significantly less than that predicted by the simple analysis and is usually in the range of 2.0 to 2.2. The relations above indicate that, qualitatively, the Limiting Drawing Ratio is:

reduced by

- a higher blank-holder force B;
- greater strain-hardening, because the rate of increase in the average flow stress in the flange will be greater than the strengthening of the cup wall;

and *increased* by

- better lubrication reducing the friction coefficient μ.;
- a more ample die corner radius, increasing the bend ratio ρ/t;
- anisotropy characterized by $R>1$.

8.3 Cup height

If a disc is drawn to a cylindrical cup, the height of the cup wall will be determined principally by the diameter of the disc. As indicated above, during drawing the flange, the outer region will tend to thicken and the top of the cup could be greater than the initial blank thickness, as illustrated in an exaggerated way in Figure 8.10.

The thinnest region will be near the base at point E where the sheet is bent and unbent over the punch corner radius. At some point mid-way up the wall, the thickness will be the same as the initial thickness. An approximate estimate of the final cup height is obtained by assuming that it consists of a circular base and cylindrical wall as shown on the right-hand side of the cup diagram and that the thickness is everywhere the same as the initial value. By equating volumes,

$$\pi R_0^2 t_0 = \pi r_i t_0 + 2\pi r_i t_0 h$$

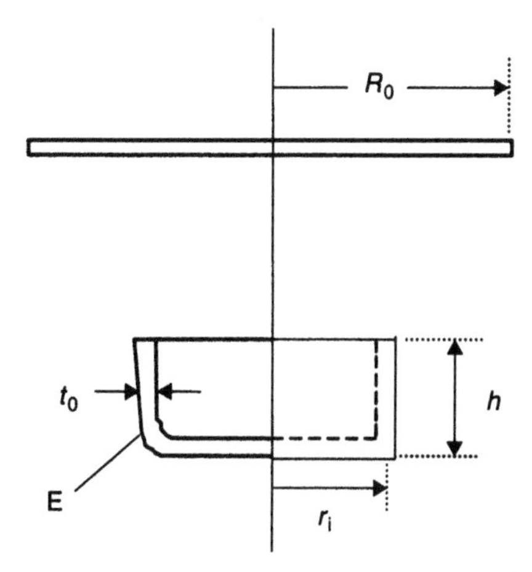

Figure 8.10 A disc of initial radius, R_0, and thickness, t_0, drawn to a cylindrical cup of height h.

and the cup height is given by

$$h \approx \frac{r_i}{2}\left\{\left(\frac{R_o}{r_i}\right)^2 - 1\right\} \tag{8.11}$$

As indicated in Section 8.2.3, the drawing ratio R_o/r_i, is usually less than about 2.2; Equation 8.11 shows that the cup height for this ratio is nearly twice the wall radius, or the height to diameter ratio of the cup is just less than unity. Deeper cups can be obtained by redrawing as described below.

8.4 Redrawing cylindrical cups

In Figure 8.11, a cup of radius r_1 and thickness t is redrawn without change in wall thickness to a smaller radius r_2. If the tension in the wall between the bottom of the punch and the die is T_ϕ, the force exerted by the punch is

$$F = 2\pi r_2 T_\phi \tag{8.12}$$

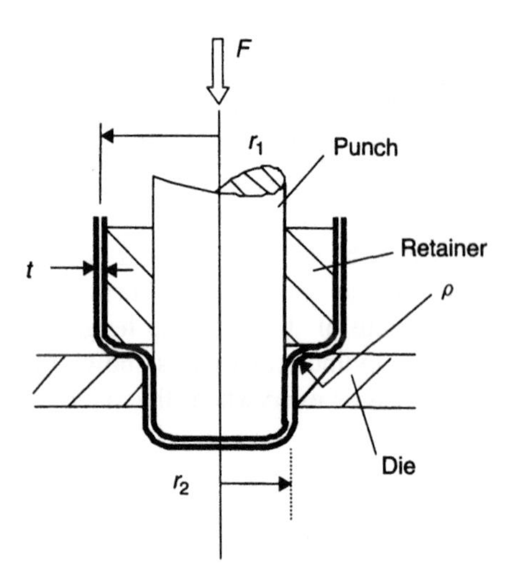

Figure 8.11 Forward redrawing of a deep-drawn cup.

Assuming that the yield tension $\overline{T}$ remains constant, then from Equation 7.10, the wall tension is

$$T_\phi = \overline{T}\ln\frac{r_1}{r_2} = \sigma_f t\ln\frac{r_1}{r_2} \tag{8.13}$$

It is shown later, in Section 10.5.1, that in plane strain bending or unbending under tension, there will be an increase in the tension given by Equation 10.20. As an approximation here, the flow stress will be substituted for the plane strain yield stress, the efficiency η taken as unity, and the term T/T_y neglected as the tension in redrawing is usually not very high. Thus for either a bend or unbend, the tension increase is

$$\Delta T_\phi \approx \frac{\sigma_f t^2}{4\rho} \tag{8.14}$$

It may be seen from Figure 8.11, that there are two bend and two unbend operations in forward redrawing. Combining Equations 8.12–8.14, the redrawing force is

$$F = 2\pi t\left(T_\phi + 4\Delta T\right) = 2\pi r_2 t\sigma_f\left(\ln\frac{r_1}{r_2} + \frac{t}{\rho}\right) \tag{8.15}$$

This shows that the redrawing force increases with larger reductions and with smaller bend ratios ρ/t.

Another form of redrawing is shown in Figure 8.12. This is reverse redrawing and the cup is turned inside out. An advantage is that there is only one bend and one unbend operation and the force is reduced to

$$F = 2\pi r_2 t\sigma_f\left(\ln\frac{r_1}{r_2} + \frac{t}{2\rho}\right) \tag{8.16}$$

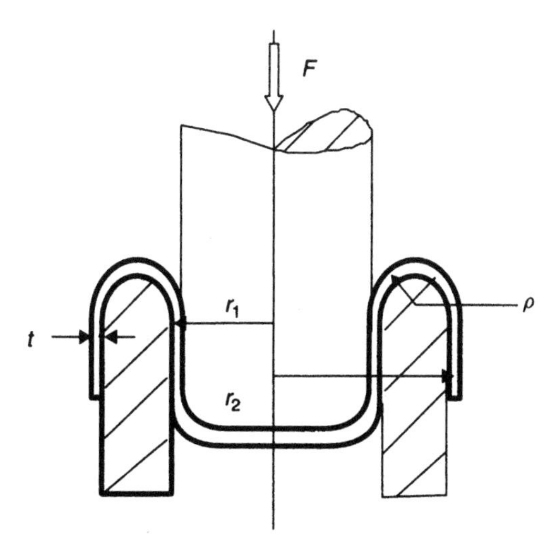

Figure 8.12 Reverse redrawing of a cylindrical cup.

Reverse redrawing can require a smaller punch force if the difference between the initial radius and the final radius is large compared with the wall thickness. If it is small, the die radius ratio ρ/t will also be small increasing the tension increase due to bending. With forward redrawing, the radii ρ can be greater than $(r_2 - r_1)/2$.

8.5 Wall ironing of deep-drawn cups

The wall of a deep-drawn cup can be reduced by passing the cup through a die, as shown in Figure 8.13. The clearance between the die and the punch is less than the initial thickness of the cup wall. As the punch must remain in contact with the base of the cup, the velocity of the material as it exits the die, v_p, is the same as the punch velocity. During ironing, there is no change in volume and the rate at which material enters the die equals the rate leaving, therefore,

$$2\pi r_i t_1 v_1 = 2\pi r_i t_2 v_p$$

or

$$v_1 = v_p \frac{t_2}{t_1} \tag{8.17}$$

The punch is thus moving faster than the incoming material and the friction force between the punch and the sheet is downwards. This assists in the process. The friction force between the die and the sheet opposes the process. It is an advantage in ironing to have a high punch-side friction μ_p, and for this reason the punch is often roughened slightly. On the other hand, the die-side friction μ_d should be low and usually the outside of the cup is flooded with lubricant.

An approximate model can be created for ironing a rigid, perfectly plastic material with a flow stress $\sigma_f = Y =$ constant. Assuming a Tresca yield condition, the through-thickness stress at entry, where the axial stress is zero, is, $-Y$, and

$$q = -\sigma_t = Y \tag{8.18}$$

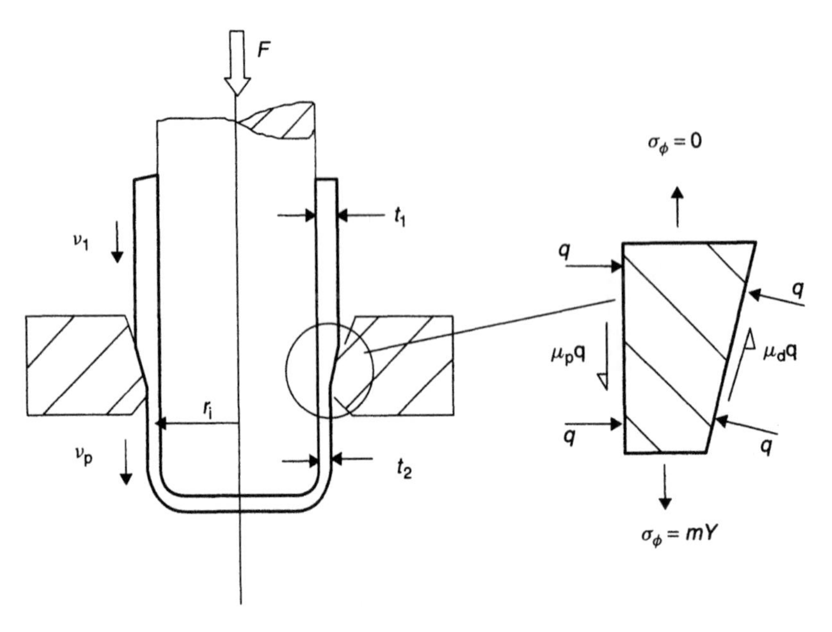

Figure 8.13 Ironing of the wall of a cylindrical deep-drawn cup.

At exit, the axial stress must be less than the yield stress to ensure that deformation occurs only within the die. As shown in Figure 8.13, this stress is mY where $m < 1$. The through-thickness stress is $\sigma_t = -(Y - mY)$, and at exit

$$q = -\sigma_t = Y(1 - m) \tag{8.19}$$

The average contact pressure is

$$\overline{q} = Y\frac{1 + (1 - m)}{2} = Y\left(1 - \frac{m}{2}\right) \tag{8.20}$$

The forces on the deformation zone are shown in Figure 8.14. The equation of equilibrium of forces in the vertical direction is

$$mYt_2 + \mu_p\overline{q}\frac{\Delta t}{\tan\gamma} = \mu_d\overline{q}\frac{\Delta t}{\sin\gamma}\cos\lambda + \overline{q}\frac{\Delta t}{\sin\lambda}\sin\gamma$$

Substituting Equation 8.20 this reduces to

$$\frac{|\Delta t|}{t_2} = \frac{2m}{2 - m} \cdot \frac{1}{\left\{1 - \dfrac{(\mu_p - \mu_d)}{\tan\gamma}\right\}} \tag{8.21}$$

The limiting condition is when $m = 1$, and the greatest thickness reduction is

$$\left(\frac{|\Delta t|}{t_2}\right)_{\max.} = \frac{1}{\left\{1 - \dfrac{(\mu_p - \mu_d)}{\tan\gamma}\right\}} \tag{8.22}$$

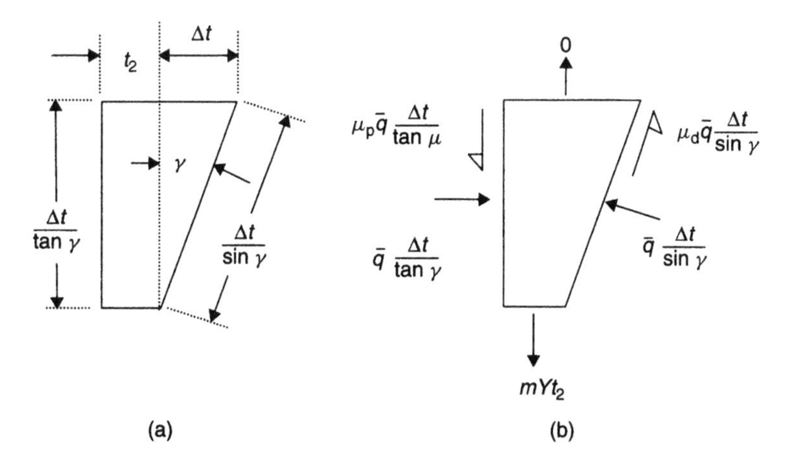

Figure 8.14 (a) Geometry of the deforming zone. (b) Forces acting on the deforming zone in ironing.

These equations give the magnitude of the thickness reduction, which is a positive quantity. It is seen that the maximum reduction is when $\mu_p > \mu_d$, i.e. when, as mentioned above, the punch is slightly rough.

Quite large reductions are possible in this process and several ironing dies may be placed in tandem. In such a case, the dies may be spaced widely apart to ensure that the wall has left one die before entering the next. This ensures that the entry stress is zero and prevents excessive wall stresses below the die. In ironing aluminium beverage cans, three dies are used and the wall thickness is reduced to about a third of the original thickness. This increases the cup height greatly.

8.6 EXERCISES

Ex. 8.1 A fully work-hardened aluminium blank of 100 mm diameter and 1.2 mm thickness has a constant flow stress of 350 MPa. It is deep-drawn in a constant thickness process to a cylindrical cup of 50 mm mid-wall diameter. The blank-holder force is 30 kN and the friction coefficient is 0.1.

(a) What is the final height?
(b) What is the approximate value of the maximum punch force?

[*Ans : (a)37.5 mm, (b)57kN.*]

Ex. 8.2 The typical characteristic between drawing stress and punch travel for a strain-hardening material was shown in Figure 8.6. Assuming the same die and punch geometry, and an efficiency of 100%, show the effect of work-hardening on the punch load by sketching two curves of drawing stress vs. punch travel for $n = 0.1$ and $n = 0.2$.
[Ans: With increasing n, the curve shifts to the right, but the maximum load decreases.]

Ex. 8.3. A sheet metal part is drawn over a die corner of radius ρ, and the frictional coefficient between the sheet and die is μ. Calculate the difference in the tensions $(T_1 - T_2)$ neglecting change in the thickness.

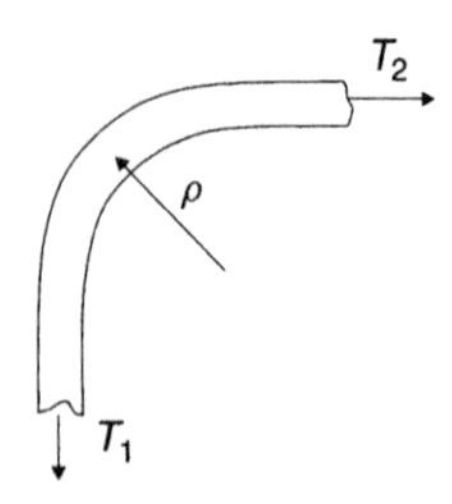

$\left[Ans : T_2 - T_1 = T_1 \left[\exp\left(-\frac{\pi}{2}\mu\right) - 1\right]\right]$

Ex. 8.4. Some years ago, a beer can had the approximate dimensions,

Diameter,	$d_f = 62\,\text{mm}$
Height,	$h_f = 133\,\text{mm}$
Wall thickness,	$t_f = 0.13\,\text{mm}$
Bottom thickness,	$t_0 = 0.41\,\text{mm}$ (initial blank thickness)
Material yield stress,	$\sigma_f = 380\,\text{MPa}$ (constant)

(a) Assuming that drawing is a constant thickness process and that the efficiency is not likely to exceed 60%. Can the cup be drawn to the final diameter in one operation?
(b) If the actual process uses two draws of equal drawing ratio. What is the intermediate diameter, d_i, and cup height, h_i?
(c) Determine the punch force in the first draw, assuming 65% efficiency.
(d) What is the cup height at the finish of the second draw?
(e) What is done to achieve the full height?
(f) Over the years, the starting thickness has been reduced to 0.33 mm. If Al alloy costs $5/kg, Al density is 2800 kg/m^3 and 100 billion cans/year are made. How much money is saved? (Assume blank diameter is unchanged.)

[Ans: (a) No; (b) 86 mm, (c) 21.6 kN, (d) 42.6 mm, (e) Ironing, (f) $1.25 Billion]

9 Stretching circular shells

9.1 Bulging with fluid pressure

9.1.1 The hydrostatic bulging test

Sheet may be bulged to an approximately spherical shape by fluid pressure as shown in Figure 9.1. This process is used to obtain mechanical properties in sheet in the so-called *hydrostatic bulging test.* As shown, the sheet is clamped rigidly around the edge.

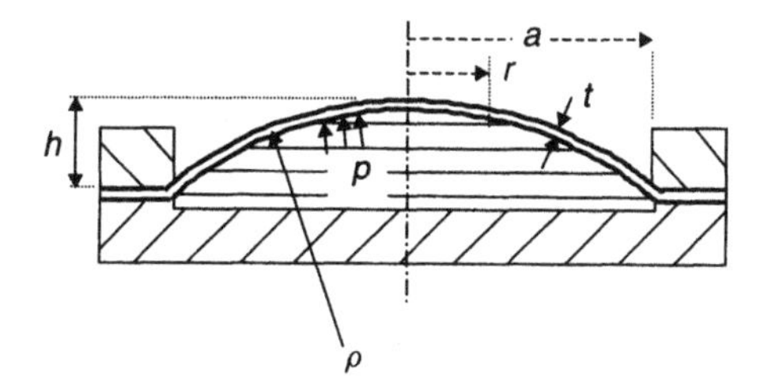

Figure 9.1 Bulging a thin sheet with fluid pressure.

The membrane stresses are illustrated in Figure 9.2(a). By symmetry, the stresses and strains at the pole are

$$\sigma_\phi; \qquad \sigma_\theta = \alpha.\sigma_\phi = \sigma_\phi; \qquad \sigma_t = 0$$

$$\varepsilon_\phi; \qquad \varepsilon_\theta = \beta.\varepsilon_\phi = \varepsilon_\phi; \qquad \varepsilon_t = \ln\frac{t}{t_0} = -2\varepsilon_\phi \tag{9.1}$$

If the sheet is deforming, from Equations 2.18(b) and 2.19(c), the effective stress and strain are

$$\bar{\sigma} = \sigma_f = \sigma_\phi \qquad \text{and} \qquad \bar{\varepsilon} = 2\varepsilon_\phi = -\varepsilon_t \tag{9.2}$$

From Equation 7.3, the stress at the pole is

$$\sigma_\phi = \sigma_\theta = \frac{p\rho}{2t} \tag{9.3}$$

As the edge of the disc is clamped, the circumferential strain at the edge must be zero, as shown in Figure 9.2(b). As indicated, the membrane strains become equal at the pole where $r = 0$.

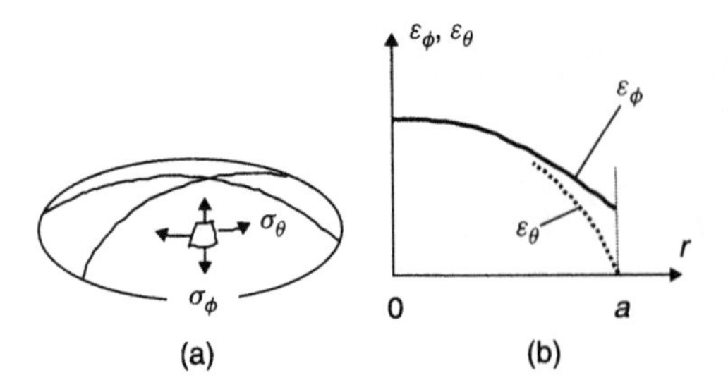

Figure 9.2 (a) Membrane stresses on the spherical shell. (b) Distribution of membrane strains at some stage in the bulging process.

As indicated, this process is used as a test for mechanical properties. The strain at the pole can be measured by a thickness gauge, or by measuring in some way the expansion of a circle at the pole; this information, together with measurement of the curvature of the pole and the current bulging pressure, permits calculation of the effective stress–strain curve from Equations 9.1 to 9.3. If the material is anisotropic, the stress–strain curve obtained shows the properties in the through-thickness direction. (It may be shown that the stress state at the pole is equivalent to hydrostatic tension plus uniaxial compression in the through-thickness direction and it is assumed that the hydrostatic stress will not influence yielding.)

The most important reason for using this test is that quite large strains can be obtained before failure even in materials having very little strain-hardening. Following Section 5.4.1, and as shown in Figure 9.3, the membrane strain at failure in biaxial stress, $\varepsilon_\phi^* = \varepsilon_\theta^*$, is greater than the strain at necking in the tensile test. Also, from Equation 9.2, the effective strain is twice the membrane strain in biaxial tension.

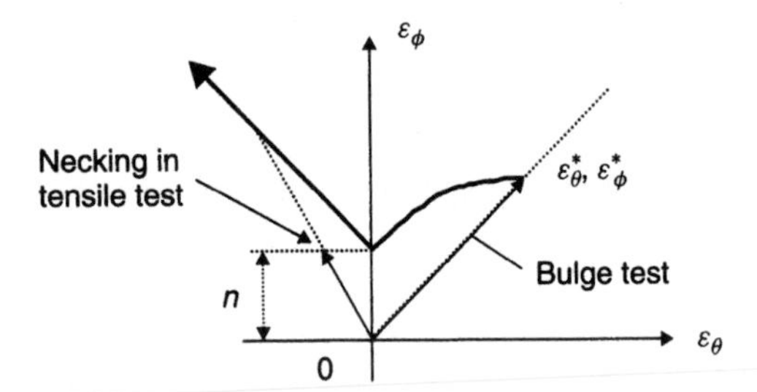

Figure 9.3 Forming limit diagram for a low-strain-hardening material showing the end points in the tensile and bulge tests.

9.1.2 An approximate model of bulging a circular diaphragm

An approximate model of the process can be obtained if it is assumed that the surface is spherical, that the membrane strains are equal everywhere and not just at the pole and that the thickness is uniform. The area of the deformed surface in Figure 9.1 is $2\pi\rho h$ and equating volumes before and after deformation gives

$$\pi a^2 t_0 = 2\pi\rho h t$$

from which

$$t = t_0 \frac{a^2}{2\rho h} \tag{9.4}$$

From Equation 9.2, for a material obeying the stress–strain law $\overline{\sigma} = K\overline{\varepsilon}^n$, the membrane stress is

$$\sigma_\phi = \overline{\sigma} = K\left(\ln \frac{t_0}{t}\right)^n$$

From Equation 9.3, the pressure to bulge the diaphragm is

$$p = \frac{2\sigma_\phi t}{\rho} = 4\overline{\sigma} t_0 \frac{h}{a^2} \frac{1}{\left\{1 + (h/a)^2\right\}^2} \tag{9.5}$$

using

$$\rho = \frac{\left(a^2 + h^2\right)}{2h} \tag{9.6}$$

In bulging a diaphragm, the pressure may reach a maximum ($dp = 0$) and then bulging continues under a falling pressure gradient. Rupture will occur when the strain at the pole reaches the forming limit as shown in Figure 9.3. This provides an example of different instabilities in processes, as discussed in Section 5.1. If the diaphragm is considered as a load-carrying structure, then the maximum pressure point constitutes instability and failure. If it is a metal forming process, instability is when necking and tearing occur at the forming limit curve, which, as mentioned, is usually beyond the maximum pressure point. As discussed in Section 5.1, most metal forming processes are displacement controlled, rather than load controlled, and local necking usually governs the end-point.

9.1.3 (Worked example) the hydrostatic bulging test

Equipment designed to obtain an effective stress strain curve by bulging a circular diaphragm with hydrostatic pressure is shown, in part, in Figure 9.4. An extensometer measures the current diameter D of a small circle near the pole of original diameter D_0 and a spherometer measures the height of the pole above this circle h. The current pressure is p and the original thickness of the sheet is t_0. Assuming that within this circle a state of uniform biaxial tension exists and that the shape is spherical, obtain relations for the effective stress and strain at this instant.

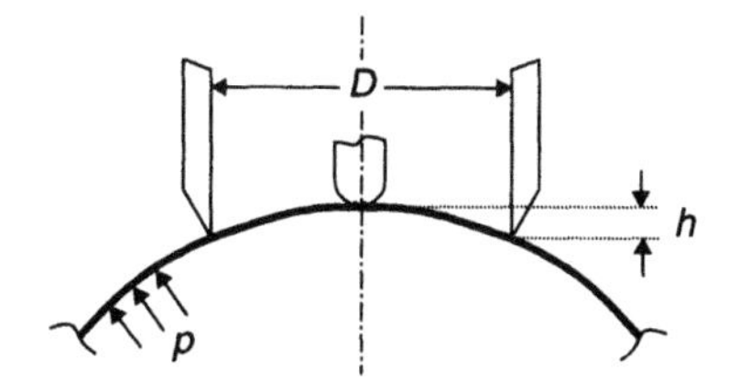

Figure 9.4 Small region at the pole in a hydraulic bulge test.

Solution. The hoop strain around a circle of diameter D is,

$$\varepsilon_\theta = \ln\frac{\pi D}{\pi D_0} = \ln\frac{D}{D_0}$$

Assuming $\varepsilon_\theta = \varepsilon_\phi$, from Equation 9.1,

$$\overline{\varepsilon} = 2\varepsilon_\theta = 2\ln(D/D_0) \qquad \text{and} \qquad t = t_0\,(D_0/D)^2$$

(Note that this is a different relation from Equation 9.4 in the approximate model, Section 9.1.2. In the approximate model, the whole of the diaphragm is considered and the dimension a is fixed. With the extensometer, the gauge circle of diameter D increases during deformation. An alternative relation for strain can be obtained assuming that the volume of the spherical cap is $2\pi\rho h t$, and that this originally was of volume $\pi D_0^2 t_0/4$. This leads to a very similar result to that above.)

The radius of curvature of the surface, from Equation 9.6, for $a = D/2$, is

$$\rho = \frac{h\left\{(D/h)^2 + 4\right\}}{8}$$

and from Equation 9.1 and 9.5,

$$\overline{\sigma} = \frac{p\rho}{2t} = \frac{p}{16}\frac{h}{t_0}\left\{(D/h)^2 + 4\right\}\left(\frac{D}{D_0}\right)^2$$

9.2 Stretching over a hemispherical punch

If a disc is clamped at the edge and stretched by a hemispherical punch, as shown in Figure 9.5, the tension in the sheet will increase with punch displacement. If there is no friction between the sheet and the punch, the greatest strain will be at the pole and the strain distribution will be similar to that in Figure 9.2. Failure would be anticipated by tearing at the pole. In practice, it is very difficult to achieve near-frictionless conditions and the effect of friction is investigated here.

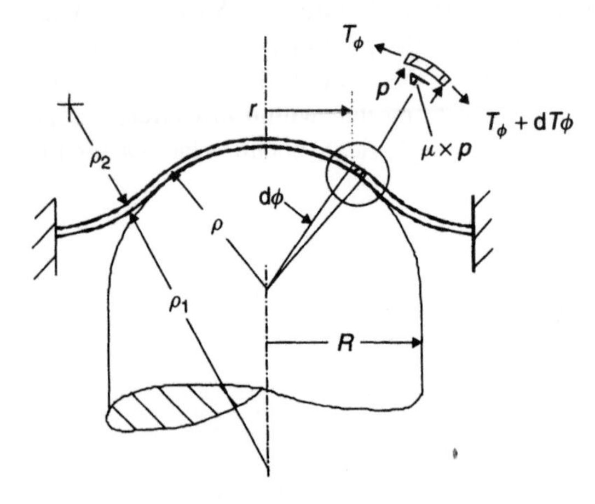

Figure 9.5 Stretching a circular blank with a hemispherical punch.

The distribution of strain, ε_ϕ, from the centre to the clamped edge is shown in Figure 9.6(a). Because of the frictional contact stress μp, the maximum tension and strain is at some distance from the pole and this is where failure by splitting is expected. The strain along a meridian is plotted in the forming limit diagram in Figure 9.6(b). The sheet tends to thin near B and then tear around a circle at B, where B is an intermediate point between the pole, A, and the edge, C.

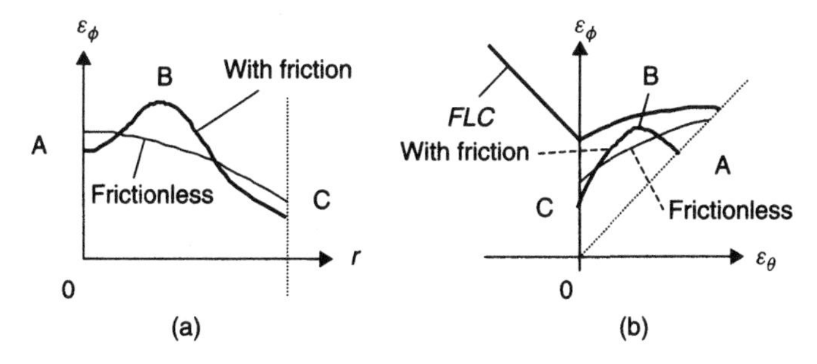

Figure 9.6 (a) Strain distribution for stretching a circular blank and (b) plot of the strains along a meridian for stretching with and without friction at the punch face.

Outside the contact region, the surface pressure is zero and from Equation 7.3 the principal curvatures are connected by

$$\frac{T_\theta}{\rho_1} = -\frac{T_\phi}{\rho_2}$$

As both T_θ and T_ϕ are tensile (positive) in punch stretching, it follows that the principal curvatures are of opposite sign and the surface is of negative Gaussian curvature as is shown in Figure 9.5.

9.2.1 (Worked example) punch stretching

In stretching a sheet over a hemispherical punch as shown in Figure 9.5, the punch diameter is 100 mm and the initial sheet thickness is 0.9 mm. The tangent point dividing the contact from the non-contact region is at a radius of 28 mm. Grid circles on the sheet initially of 3.5 mm diameter are measured at the tangent point; along the meridional direction the major axis of the deformed circle is 4.4 mm and the minor axis in the hoop direction is 4.1 mm. The material has a stress strain relation of $\overline{\sigma} = 700(\overline{\varepsilon})^{0.2}$ MPa. Determine the punch force.

Solution. The major and minor strains and the strain ratio are

$$\varepsilon_\phi = \ln(4.4/3.5) = 0.229; \quad \varepsilon_\theta = \ln(4.1/3.5) = 0.158 \quad \text{and}$$

$$\beta = (0.158/0.229) = 0.69$$

The thickness strain is

$$\varepsilon_t = -(1 + 0.69) \times 0.229 = -0.387$$

and the current thickness is

$$t = t_0 \exp(-\varepsilon_t) = 0.9 \times \exp(-0.387) = 0.61\,\text{mm}$$

The stress ratio, from Equation 2.14, is

$$\alpha = \frac{2 \times 0.69 + 1}{2 + 0.69} = 0.89$$

The effective strain from Equation 2.19 is

$$\bar{\varepsilon} = \sqrt{(4/3)\left\{1 + 0.69 + 0.69^2\right\}0.229} = 0.389$$

The effective stress is

$$\bar{\sigma} = 700 \times 0.389^{0.2} = 580\,\text{MPa}$$

From Equation 2.18, the meridional stress is

$$\sigma_\phi = 580/\sqrt{1 - 0.89 + 0.89^2} = 611\,\text{MPa}$$

The meridional tension is,

$$T_\phi = 611 \times 10^6 \times 0.61 \times 10^{-3} = 373\,\text{kN/m}$$

The semi-angle subtended by the tangent circle is given by

$$\sin\phi = 28/50$$

The punch force is

$$F = 2\pi r T_\phi \sin\phi = 2\pi \times 28 \times 10^3 \times 373\,(28/50) = 37\,\text{kN}$$

9.3 Effect of punch shape and friction

For punches that are not hemispherical, the strain distribution depends on the punch shape. Two cases are shown in Figure 9.7; the amount of thinning, given by the magnitude of the thickness strain $|\varepsilon_t|$, is greatest where the curvature of the profile is greatest. Friction also affects the strain distribution and the greatest punch displacement. On the left of each diagram, the frictionless case is shown; on the right, the case where there is friction between the punch and the sheet is illustrated.

In the flat-bottomed punch, Figure 9.7(a), friction is confined to the punch corner radius. This prevents the tension across the face of the punch from increasing sufficiently to stretch the material over the face of the punch. The maximum depth that can be formed is therefore less with friction than without it. With the pointed punch, Figure 9.7(b), the reverse effect occurs. Without friction, strain is concentrated near the nose and the depth before failure is limited. The effect of friction is to reduce the tension at the nose and spread the strain over a greater area; this permits greater depth of forming.

Stretching over a punch is often a pre-form operation, where material is redistributed to obtain favourable conditions for the final stage. In the second stage, the shape of the punch is defined by the part design, but for the preform punch there is some flexibility in

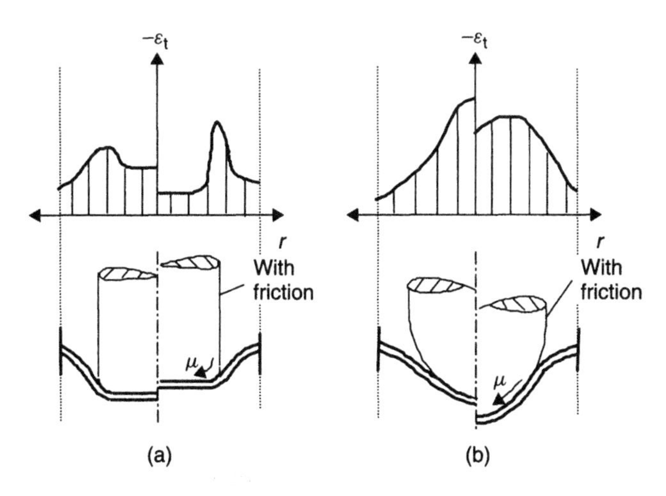

Figure 9.7 Strain distributions for (a) a flat-bottomed punch, and (b) a pointed punch, without and with friction.

the shape and the tool designer needs to consider both curvature and friction in arriving at a suitable tool.

9.4 Exercises

Ex. 9.1 In the example in Section 9.1.3, the extensometer initially rests on a circle of 50 mm diameter on the flat sheet. The initial sheet thickness is 1.2 mm. At some instant in the test, the pressure is 6.4 MPa, the spherometer measures a vertical distance of 3 mm and the extensometer indicates that the circle has grown to a diameter of 61 mm. Determine the effective stress and strain at this instant.
[Ans: 620 MPa, 0.4]

Ex. 9.2 Determine the principal radius of curvature in the meridional direction in the unsupported sheet adjacent to the tangent point in the example given in the example in Section 9.2.1.
[Ans: −56 mm)]

Ex. 9.3 Using an approximate analysis, determine the pressure versus bulge height characteristic for the operation shown in Figure 9.1. A disk of 1.2 mm thickness is clamped around a circle of 100 mm diameter and bulged to a height of 45 mm. The stress–strain curve is $\overline{\sigma} = 350\overline{\varepsilon}^{0.18}$ MPa. Determine the effective strain at maximum pressure.
[Ans: (0.4)]

10 Combined bending and tension of sheet

10.1 Introduction

In Chapter 6, the bending of sheet under a pure moment was studied. Here we investigate several situations in which both a tension and a moment are applied to the sheet. In the first instance, we consider an elastic, perfectly plastic sheet bent elastically over a former and then tension is applied. Combined tension and moment on a rigid, perfectly plastic sheet is then analysed and the case of sheet being dragged under tension over some die radius is studied. A similar nomenclature to that given in Section 6.2. is used; i.e. continuous sheet is subject to a force per unit width, or tension T applied at the mid-surface and moment per unit width of M. Bending occurs under a plane stress, plane strain state; the elastic modulus is E' as given in Equation 6.10, and the plane strain yield stress is S. It should be noted that because the analysis of plastic bending is non-linear, the order in which the tension and the moment are applied may influence the result.

10.2 Stretching and bending an elastic, perfectly plastic sheet

If the desired curvature of a sheet is less than the limiting elastic curvature, the sheet cannot be formed to shape simply by bending over a die block or former. It would either springback to the flat shape, or, if it was over-bent until it became partially plastic, the springback would be so great that the process would be difficult to control. In gently curved parts such as aircraft skin panels, the sheet is gripped on either side and pulled; it is then wrapped under tension over a former. Here we consider a case more akin to a stamping operation where the sheet is first curved elastically to the shape of the former and then tension is applied. This is illustrated in Figure 10.1, and a model is developed for a two-dimensional, frictionless case.

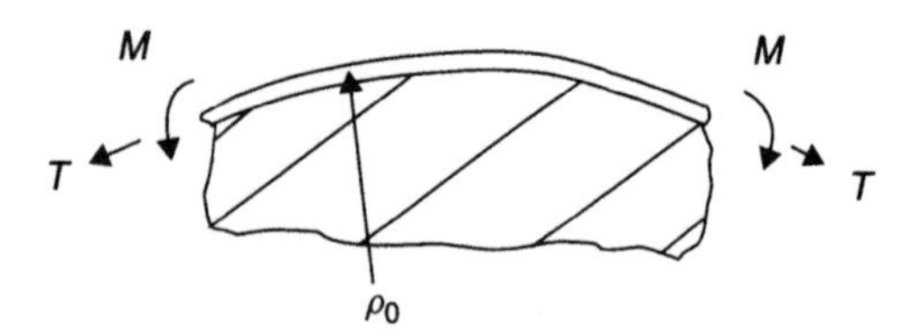

Figure 10.1 Bending and then stretching a sheet over a large radius of curvature former.

Initially when the moment is applied without any tension, the stress distribution will be as shown in Figure 10.2(a). As the tension is zero, the strain at some distance y from the mid-surface, from Equations 6.3 and 6.4, is

$$\varepsilon_1 = \frac{y}{\rho_0}$$

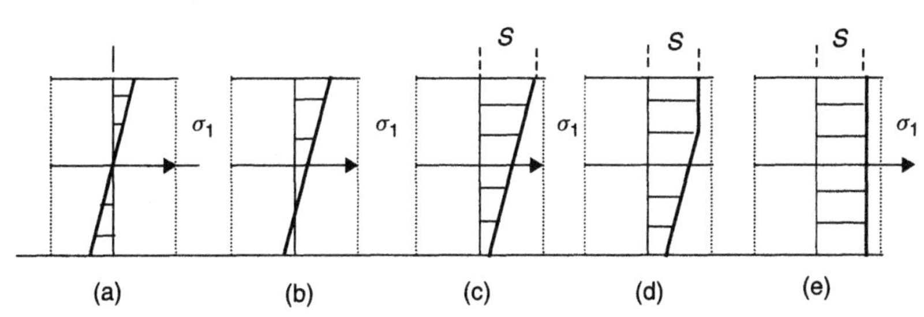

Figure 10.2 Development of the stress distribution as the tension in the sheet increases.

and the stress, as the material is in the elastic state, is

$$\sigma_1 = E' \frac{y}{\rho_0}$$

The slope of the stress distribution in Figure 10.2(a) is

$$\frac{d\sigma_1}{dy} = \frac{E'}{\rho_0}$$

and this remains constant as the tension increases as shown in Figures 10.2(b) and (c). The sheet will start to become plastic when the stress at the outer fibre reaches the yield stress S, in (c). With further tension, a zone of plastic deformation increases from the outside (d), until the whole section is plastic (e). The tension at this point is

$$T_y = St \tag{10.1}$$

When the tension is equal to the yield tension, the stress distribution is uniform and the moment is zero. The initial moment, when $T = 0$, is, from Equation 6.16,

$$M = \frac{E't^3}{12}\left(\frac{1}{\rho_0}\right) = M_0 \tag{10.2}$$

and this remains constant until the sheet just starts to yield, as in (c).

At any point in the elastic stage, between (a) and (c) in Figure 10.2, the stress distribution can be considered as the sum of a uniform stress equal to that at the middle surface, σ_{1a}, plus a constant bending stress distribution as shown in (a); i.e. from Equation 6.17, the bending stress is

$$\sigma_{1b} = E'\left(\frac{1}{\rho_0}\right)y \tag{10.3}$$

At the limiting case (c), the outer fibre stress is S, so the stress at the middle surface is

$$(\sigma_1)_{y=0} = S - E'\left(\frac{1}{\rho_0}\right)\frac{t}{2}$$

and the tension on the section is

$$T = (\sigma_1)_{y=0}\, t = St - E'\left(\frac{1}{\rho_0}\right)\frac{t_2}{2} = T_y\left(1 - \frac{\rho_e}{\rho_0}\right) \tag{10.4}$$

where $T_y = St$, is the tension required to yield the sheet in the absence of any moment. The limiting elastic radius of curvature ρ_e is given by Equation 6.19. Equation 10.4 shows that if the radius of curvature of the former, ρ_0, is very large compared with the limiting elastic radius of curvature ρ_e, the applied tension will nearly reach the yield tension T_y before the sheet starts to yield plastically.

When the tension exceeds this value, the moment will start to reduce. At some instant, the elastic, plastic interface is at a distance $mt/2$ from the middle surface as shown in Figure 10.3, where $-1 < m < 1$. At this level, the strain is equal to the yield strain S/E', and following Equations 6.3 and 6.4, the strain is given by

$$\varepsilon_1 = \varepsilon_a + y\left(\frac{1}{\rho_0}\right)$$

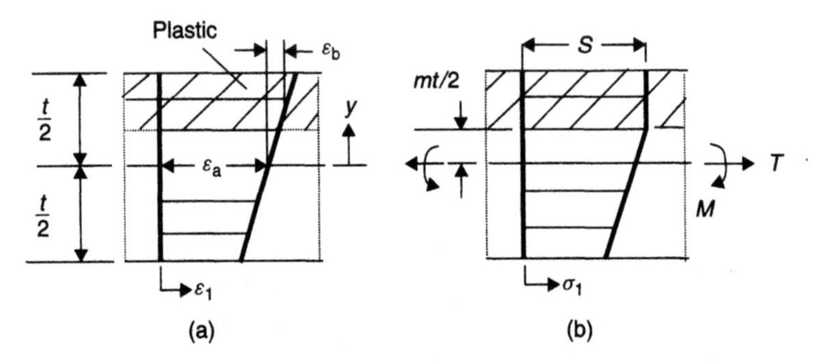

Figure 10.3 Distribution of strain (a) and stress (b) in an elastic, perfectly plastic sheet bent to a gentle curvature and stretched.

Substituting $\varepsilon_1 = S/E'$, at $y = mt/2$, we obtain

$$\varepsilon_a = \frac{S}{E'} - \frac{1}{\rho_0}\frac{mt}{2}$$

The strain at a distance y from the middle surface is therefore

$$\varepsilon_1 = \varepsilon_a + \varepsilon_b = \frac{S}{E'} - \frac{1}{\rho_0}\frac{mt}{2} + \frac{y}{\rho_0}$$

The stress distribution shown in Figure 10.3 in the elastic range is

$$\sigma_1 = E'(\varepsilon_a + \varepsilon_b) = S - \frac{E'}{\rho_0}\left(\frac{mt}{2} - y\right) \tag{10.5}$$

From Equation 6.8, the equilibrium equation is

$$M = \int_{-t/2}^{mt/2} \left\{S - \frac{E'}{\rho_0}\left(\frac{mt}{2} - y\right)\right\} y\, dy + \int_{mt/2}^{t/2} Sy\, dy$$

This reduces to

$$M = \frac{E't^3}{12\rho_0}\left(\frac{2+3\ m-m^3}{4}\right) = M_0\left(\frac{2+3\ m-m^3}{4}\right) \tag{10.6}$$

The tension in the elastic, plastic state is, from Equation 6.7,

$$T = \int_{-t/2}^{mt/2}\left\{S - \frac{E'}{\rho_0}\left(\frac{mt}{2}-y\right)\right\}\mathrm{d}y + \int_{mt/2}^{t/2} S\,\mathrm{d}y$$

which reduces to

$$T = T_{\mathrm{y}}\left\{1 - \frac{1}{4}\frac{\rho_{\mathrm{e}}}{\rho_0}(m+1)^2\right\} \tag{10.7}$$

The relation between moment and applied tension is shown schematically in Figure 10.4(a). Assuming that the unloading process is elastic, Equation 6.30 shows that the springback is proportional to the change in moment. If the moment in the loaded condition is reduced to zero by applying a yielding tension, the springback on unloading will be zero. Thus a yielding tension will 'set' the shape in the sheet to that of the former. In an actual forming operation, the sheet will be stretched to obtain a small plastic strain throughout; this will ensure that negligible springback occurs.

In forming operations of gently curved panels, the required surface may be doubly-curved rather than cylindrical as examined here. The problem of springback is then more complicated than in the simple model, but this model does illustrate the principle that tension, if it is large enough, will reduce bending springback significantly. Although not shown, tension will also reduce residual stresses as described in Section 6.6.2 and this will enhance the structural performance of the panel.

10.2.1 (Worked example) curving an elastic, perfectly plastic sheet

(a) A form block, as shown in Figure 10.1, is 1.4 m wide and has a crown height of 105 mm. A steel sheet, 1.5 mm thick is bent over this block and then stretched. For the sheet, the Youngs Modulus is 200 GPa, Poissons ratio is 0.3, the flow stress in uniaxial tension is 220 MPa, and the material may be considered as non-work-hardening. Assuming that the block is frictionless, determine the moment in the sheet as the tension is increased up to fully plastic yielding.

Solution. If a is the half-width of the block and h the crown height, the radius of curvature is

$$\rho_0 = \frac{a^2+h^2}{2h} = \left(0.7^2+0.105^2\right)/2 \times 0.105 = 2.39\ \mathrm{m}$$

The plane strain elastic modulus is,

$$E' = \frac{E}{1-\upsilon^2} = \frac{200}{1-0.3^2} = 220\ \mathrm{GPa}$$

The plane strain yield stress is

$$S = \left(2/\sqrt{3}\right) \times 220 = 254\,\mathrm{MPa}$$

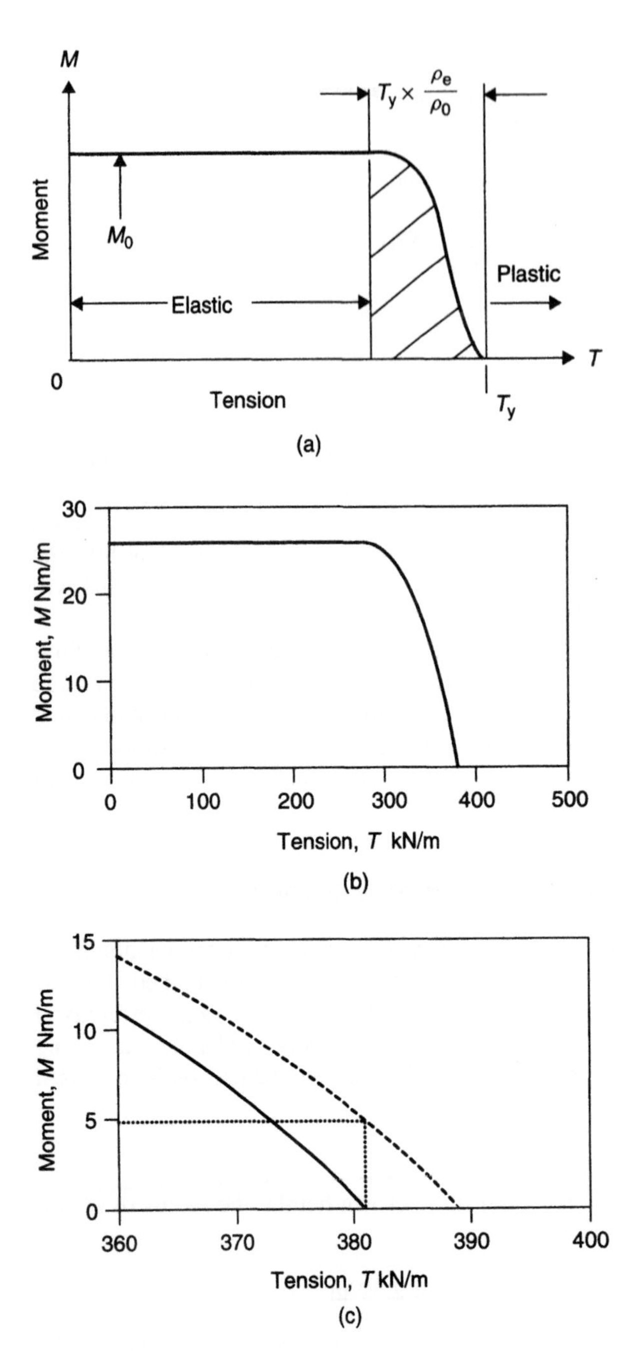

Figure 10.4 (a) Effect of tension on the moment on an elastic, perfectly plastic sheet. (b) Moment versus tension for the worked example in Section 10.2.1. (c) Moment tension relation for materials with different flow stresses. For the upper curve, the flow stress is 2% greater than that for the lower curve.

The limiting elastic radius of curvature, from Equation 6.19, is

$$\rho_e = E't/2S = 220 \times 10^9 \times 1.5 \times 10^{-3}/2 \times 254 \times 10^6 = 0.650 \text{ m}$$

The elastic moment in the sheet as it is bent over the former, from Equation 10.2, is

$$M_o = \frac{E't^3}{12}\frac{1}{\rho_0} = \frac{220 \times 10^9 \times \left(1.5 \times 10^{-3}\right)^3}{12 \times 2.39} = 25.9 \text{ Nm / m}$$

The fully plastic yielding tension is

$$T_y = St = 254 \times 10^6 \times 1.5 \times 10^{-3} = 381 \text{ kN / m}$$

The moment in the sheet as the tension is increased, from Equation 10.6, is

$$M = 25.9\left(\frac{1}{2} + \frac{3m}{4} - \frac{m^3}{4}\right) \text{ Nm / m}$$

The tension in the sheet is given as a function of m by Equation 10.7, i.e.

$$T = 381\left\{1 - \frac{0.650}{4 \times 2.39}(m+1)^2\right\}$$

Computing the tension and moment in the range $-1 \leq m \leq 1$, one obtains the characteristic shown in Figure 10.4(b). We note that the moment remains constant in the elastic range, and then falls rapidly to zero as the tension increases to the yield tension.

(b) Consider the case in which the above process is controlled by setting the tension to the yield tension, 381 kN/m. Imagine that new sheet is formed with all conditions the same, except that the yield stress of the sheet is 2% greater; i.e. the yielding tension for the sheet is 389 kN/m. If the tension is not changed, there will be some residual moment. Determine approximately the final radius of curvature of the new sheet.

Solution. The new plane strain yield stress is

$$S = 1.02 \times 254 = 259 \text{ MPa}$$

and the yield tension is

$$T_y = 259 \times 10^6 \times 1.5 \times 10^{-3} = 389 \text{ kN / m}$$

The limiting elastic radius of curvature is

$$\rho_e = 220 \times 10^9 \times 1.5 \times 10^{-3}/2 \times 259 \times 10^6 = 0.637$$

The tension during stretching is,

$$T = 389\left\{1 - \frac{0.637}{4 \times 2.39}(m+1)^2\right\} \text{ kN / m}$$

In computing a new characteristic, only the tension is changed from the above. The characteristics for the two materials are shown in an enlarged view in Figure 10.4(c).

For a tension of 381 kN/m, the moment in the sheet is approximately 4.8 Nm/m. For elastic unloading, the change in curvature is proportional to the change in moment, and from Equation 6.30,

$$\Delta\left(\frac{1}{\rho}\right) = \frac{\Delta M}{E'I} = \frac{4.8}{220 \times 10^9 (1.5 \times 10^{-3})^3/12} = 0.078\,\mathrm{m}^{-1}$$

The curvature of the form block is $1/2.39 = 0.418\,\mathrm{m}^{-1}$, hence the final curvature is

$$0.418 - 0.078 = 0.340\,\mathrm{m}^{-1}$$

i.e. the final radius of curvature is $1/0.340 = 2.94$, or a change in radius from that of the form block of ((2.94-2.39)/2.39)/100 = 23%. This illustrates that the curvature of the sheet is very sensitive to tension if the process is in the elastic, plastic region. For this reason, the sheet is usually overstretched, as mentioned above, to ensure that changes in the strength or thickness of the incoming sheet will not result in springback.

10.3 Bending and stretching a strain-hardening sheet

The previous section assumed that the material did not strain-harden. In practice, sheet having some strain-hardening potential will be used to ensure that it can be stretched beyond the elastic limit in a stable manner. The strain distribution will be similar to that shown in Figure 10.3, but once the whole section has yielded, the stress distribution will not be uniform; the stress will increase from the inner to the outer surface as shown in Figure 10.5. As this stress is not constant, there will be some moment in equilibrium with the section as well as the tension.

The stress–strain curve for the material is shown in Figure 10.5(a). The range of strain across the section is from just below the mid-surface strain ε_{1a}, to just above it; the stress–strain relation is assumed to be linear in this range with a slope of $|d\sigma/d\varepsilon|_{\varepsilon_{1a}}$.

The stress and strain distributions are shown in Figure 10.5. The strain is given by

$$\varepsilon_1 = \varepsilon_a + \varepsilon_b = \varepsilon_a + \frac{y}{\rho_0} \tag{10.8}$$

The stress is given by

$$\sigma_1 = \sigma_a + \frac{d\sigma_1}{d\varepsilon_1}\bigg|_{\varepsilon_{1a}} \frac{y}{\rho_0} \tag{10.9}$$

The moment associated with the stress distribution is

$$M = \int_{-t/2}^{t/2} \left\{ \sigma_a + \frac{d\sigma_1}{d\varepsilon_1}\bigg|_{\varepsilon_{1a}} \cdot \frac{y}{\rho_0} \right\} y\,dy = \frac{d\sigma_1/d\varepsilon_1}{\rho_0} \frac{t^3}{12} \tag{10.10}$$

On unloading the tension, the sheet will shrink slightly along its middle-surface, but in the two-dimensional, cylindrical case, this will only have a very small effect on the curvature. If the moment unloading process is elastic, Equation 6.30 will give the change in curvature associated with a change in moment of $-M$. As, for unit width of sheet, $I = t^3/12$, the proportional change in curvature is

$$\frac{\Delta(1/\rho)}{1/\rho_0} = -\frac{|d\sigma_1/d\varepsilon_1|_{\varepsilon_{1a}}}{E'} \tag{10.11}$$

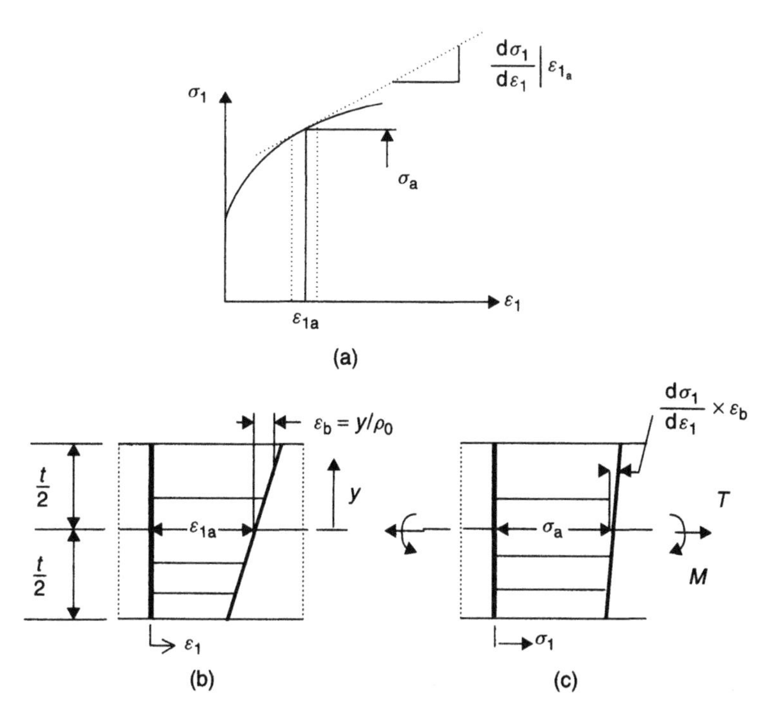

Figure 10.5 (a) Stress–strain curve for a strain-hardening material showing the range of stress in a sheet bent and stretched to a mid-surface strain of ε_{1a}. (b) The strain distribution when bent over a former of radius ρ_0, and (c) the stress distribution.

The term $d\sigma_1/d\varepsilon_1$ is the *plastic modulus*. It is much smaller than the elastic modulus, but in a material obeying a power law strain-hardening model the slope of the curve is

$$\frac{d\sigma_1}{d\varepsilon_1} = n\frac{\sigma_1}{\varepsilon_1} \tag{10.12}$$

At small strains, the slope is greatest and springback could be significant. For this reason, it is preferable to stretch a strain-hardening sheet by a few per cent so that the slope of the stress-strain curve is reduced and springback is less sensitive to strain.

10.3.1 (Worked example) curving a strain-hardening sheet

A steel sheet has a plane strain stress strain curve fitted by

$$\sigma_1 = 800\,(0.0015 + \varepsilon_1)^{0.22}\ \text{MPa}$$

It is stretched over a form block as shown in Figure 10.1. The block is 1.6 m wide and the radius of curvature is 5 m. Determine the springback in terms of change in crown height if the sheet is stretched to a mean plastic strain of (a) 0.003 (0.3%), and (b) 0.015 (1.5%). ($E' = 220$ GPa.)

Solution. The slope of the stress strain curve is

$$\frac{d\sigma_1}{d\varepsilon_1} = \frac{0.22 \times 800}{(0.0015 + \varepsilon_1)^{0.78}}\ \text{MPa}$$

The change in curvature is given by Equation 10.11, i.e.

$$\Delta (1/\rho) = -\frac{d\sigma_1}{E' d\varepsilon_1} (1/\rho_0)$$

Given that

$$\rho = \frac{a^2 + h^2}{2h} \approx \frac{a^2}{2h}$$

for small values of h/a, we obtain

$$\Delta h = \frac{a^2}{2} \Delta (1/\rho)$$

Hence

$$\Delta h = \frac{0.8^2}{2} \frac{0.22 \times 800 \times 10^6}{220 \times 10^9} \frac{1}{(0.0015 + \varepsilon_1)^{0.78}} \frac{1}{5} \mathrm{m}$$

Evaluating, for 0.3% strain, the springback is 3.5 mm and for 1.5% strain it is 1.3 mm.

10.4 Bending a rigid, perfectly plastic sheet under tension

If the bend radius is in the range of about 3 to 10 times the sheet thickness, it is reasonable in an approximate analysis to consider the material as rigid, perfectly plastic, as discussed in Section 6.5.2; also we neglect through-thickness stress and assume plane strain bending. (For very small radius bends, these assumptions may not be justified.)

If a rigid, perfectly plastic sheet is subjected to a tension less than the yield tension and then to a moment sufficient to generate some curvature, the strain and stress distributions will be as shown in Figure 10.6.

The neutral axis will be at some distance e from the mid-surface. For a positive tension the neutral surface is below the mid-surface and if the sheet is in compression, i.e. T is negative, it is above the mid-surface. The strain in Figure 10.6(b), from Equation 6.3 and 6.4, is

$$\varepsilon_1 = \varepsilon_a + \varepsilon_b = \frac{e}{\rho} + \frac{y}{\rho} \tag{10.13}$$

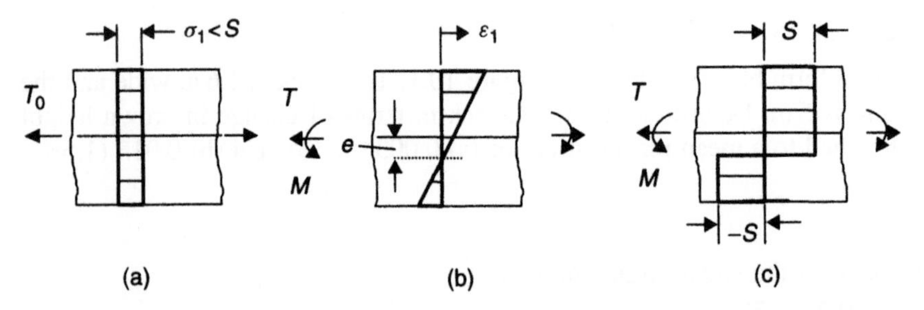

Figure 10.6 (a) Stress distribution in a sheet prior to bending. (b) Strain distribution in the sheet after bending. (c) Stress distribution after bending.

Applying the equilibrium equation, Equation 6.7, we obtain

$$T = \int_{-t/2}^{-e} -S\,dy + \int_{-e}^{t/2} S\,dy = 2Se$$

or

$$e = \frac{T}{2S} = \frac{t}{2}\frac{T}{T_y} \tag{10.14}$$

where the tension to yield the sheet in the absence of tension is $T_y = St$. The moment equilibrium equation, Equation 6.8, gives

$$M = \int_{-t/2}^{-e} -Sy\,dy + \int_{-e}^{t/2} Sy\,dy = S\left[\left(\frac{t}{2}\right)^2 - e^2\right]$$

and substituting the plastic moment M_p in the absence of tension as given by Equation 6.21, we obtain for combined tension and moment, that the moment is

$$M = \frac{St^2}{4}\left[1 - \left(\frac{T}{T_y}\right)^2\right] = M_p\left[1 - \left(\frac{T}{T_y}\right)^2\right] \tag{10.15}$$

It is seen that the presence of an axial force on the sheet will significantly reduce the moment required to bend the sheet and this will be true for both tensile and compressive forces as the sign of the tension T in Equation 10.15 will be immaterial, although, as indicated, the position of the neutral axis in Figure 10.6 does depend on the sign of the applied force.

10.5 Bending and unbending under tension

A frequent operation in sheet forming is dragging sheet over a radius as illustrated in Figure 10.7. The sheet moves to the right and is suddenly bent at A. It then slides against friction over the radius and is unbent at B. An example of such a process is shown in draw die forming in Figure 4.3. At the region DC the sheet slides against friction over the die radius. This tension is sufficient to yield the sheet and the process can only be performed with strain-hardening material. Referring to Figure 4.3, there are three important effects:

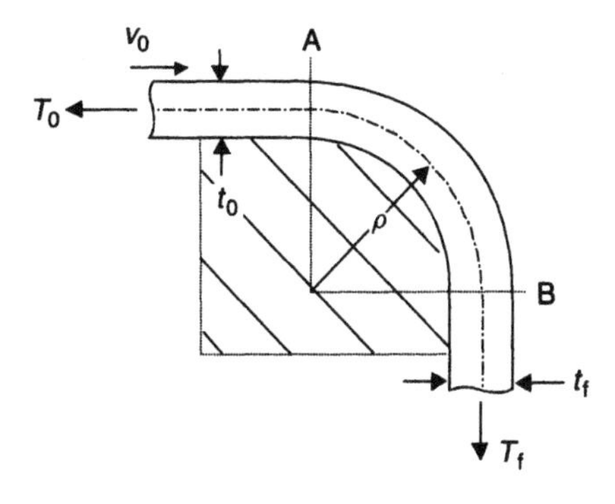

Figure 10.7 Dragging a sheet over a tool radius under tension.

- as the sheet bends at D and unbends at C it will be strain-hardened and there will be a sudden increase in tension;
- the tension will increase as the sheet slides against friction between D and C;
- at each bend or unbend, there will be a sudden reduction in sheet thickness.

All these effects will reduce the ultimate load-carrying capacity of the side-wall CB, and could affect the whole process.

The analysis of such an operation is beyond the present scope, however, a simple model can be constructed for the process shown in Figure 10.7 for the case in which the tension in the sheet is lower than that required to yield it in tension alone and where we assume rigid, perfectly plastic behaviour for the sheet. The model is an approximate one and follows that given in Section 10.4.

10.5.1 Bending at point A

If the tension in the sheet is rather less than the yield tension, i.e. $T_0 < St_0 = T_y$, bending will occur at a moment given by Equation 10.15. In Figure 10.8(a), an element of length $\mathrm{d}l$ entering at A is bent by an angle $\mathrm{d}\theta$ as shown in Figure 10.8(b), and as $\mathrm{d}l = \rho\,\mathrm{d}\theta$, this angle is $\mathrm{d}\theta = \mathrm{d}l/\rho$.

The plastic work done on the element by bending is

$$W_b = M\,\mathrm{d}\theta = M_p\left[1 - \left(\frac{T}{T_y}\right)^2\right]\frac{\mathrm{d}l}{\rho}$$

The time taken for the element to pass the point A is $\mathrm{d}l/v_0$, hence the rate of doing plastic work is

$$\dot{W}_b = \frac{M_p v_0}{\rho}\left\{1 - \left(\frac{T}{T_y}\right)^2\right\}$$

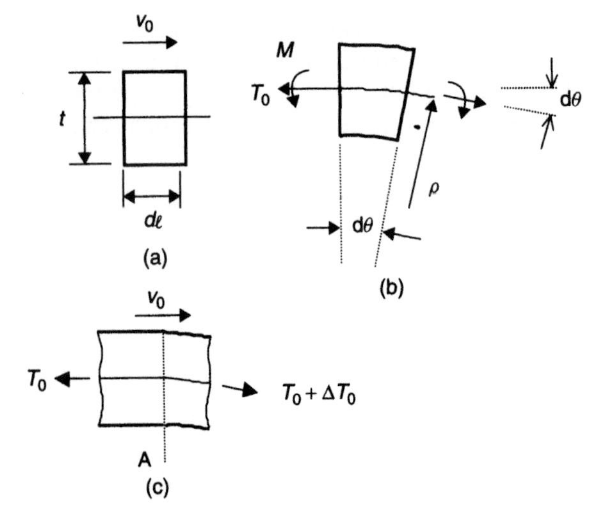

Figure 10.8 Bending of an element at point A in Figure 10.7. (a) Incoming element. (b) Element after bending at A. (c) Increase in tension at A.

Substituting for M_p from Equation 6.21, this may be written as

$$\dot{W}_b = \frac{St^2 v_0}{4\rho}\left\{1 - \left(\frac{T}{T_y}\right)^2\right\} \tag{10.16}$$

As may be seen from Figure 10.6 and Equation 10.13, the strain at the mid-surface is

$$\varepsilon_a = \frac{e}{\rho} = \frac{t}{2\rho}\left(\frac{T_0}{T_y}\right)$$

This is the average strain for the element shown in Figure 10.8(a) and therefore during bending it will extend an amount

$$\delta = \varepsilon_a \mathrm{d}l = \frac{t}{2\rho}\left(\frac{T_0}{T_y}\right)\mathrm{d}l$$

The tension will therefore do work on the element of magnitude

$$W_t = T_0\delta = \frac{t}{2\rho}\frac{T_0^2}{T_y}.\mathrm{d}l$$

As indicated, this is done in a time $\mathrm{d}l/v_0$, therefore the rate of doing tensile work is

$$\dot{W}_t = \frac{W_t}{\mathrm{d}l/v_0} = \frac{t.v_0}{2\rho}.\frac{T_0^2}{T_y} = \frac{St^2 v_0}{2\rho}\left(\frac{T_0}{T_y}\right)^2 \tag{10.17}$$

Combining Equations 10.16 and 10.17, the total rate of doing plastic work on the element is

$$\dot{W}_{pl.} = \frac{St^2 v_0}{4\rho}\left[1 + \left(\frac{T_0}{T_y}\right)^2\right] \tag{10.18}$$

It is appropriate to consider that the work done at the bend at A will be greater than that dissipated by plastic work alone and to assign an efficiency factor, $\eta < 1$, to account for friction and redundant work. The work balance may be written as

$$\dot{W}_{ext.} = \frac{\dot{W}_{pl.}}{\eta}$$

We assume that the external work done arises from a sudden increase in the tension as shown in Figure 10.8(c). This external work done at this point is

$$\dot{W}_{ext.} = [(T_0 + \Delta T_0) - T_0]\, v_0 = \Delta T_0 v_0 \tag{10.19}$$

and from Equations 10.18, 10.19 and 6.21, we obtain

$$\Delta T_0 = \frac{M_p}{\eta\rho}\left[1 + \left(\frac{T_0}{T_y}\right)^2\right] = \frac{St^2}{4\eta\rho}\left[1 + \left(\frac{T_0}{T_y}\right)^2\right]$$

$$= \frac{1}{4\eta}\frac{T_y}{(\rho/t_0)}\left[1 + \left(\frac{T_0}{T_y}\right)^2\right] \tag{10.20}$$

It has already been stated that this analysis is an approximate one that applies only when the neutral axis lies within the sheet and the tension is less than the yielding tension. Nevertheless, it is a useful relation as it shows that the increase in tension is inversely proportional to the bend ratio and increases with the back tension.

10.5.2 Thickness change during bending

As indicated above, the average strain in the sheet at the point of bending A is ε_a. As this is a plane strain process, the thickness strain will be, $\varepsilon_t = -\varepsilon_a$; for small strains we may write

$$\frac{\Delta t}{t} \approx \frac{dt}{t} \approx \varepsilon_t = -\varepsilon_a = -\frac{t_0}{2\rho}\left(\frac{T_0}{T_y}\right) = \frac{1}{2(\rho/t_0)}\left(\frac{T_0}{T_y}\right) \tag{10.21}$$

Thus there is a thickness reduction which increases linearly with tension and is greater for small bend ratios.

10.5.3 Friction between the points A and B

An element sliding on the surface between A and B in Figure 10.7 is shown in Figure 10.9. If the thickness of the sheet is small compared with the radius of the mid-surface, the contact pressure, following Section 4.2.5, is

$$p = \frac{T}{\rho} \tag{10.22}$$

The equation of equilibrium in the circumferential direction is

$$T + dT = T + \mu p \, d\theta 1$$

i.e.

$$\frac{dT}{T} = \mu \, d\theta \tag{10.23}$$

If the tension in the sheet after bending at A is $T_A = T_0 + \Delta T_0$, and before unbending at B is T_B, then integrating Equation 10.22 gives

$$[\ln T]_{T_A}^{T_B} = [\mu\theta]_0^{\pi/2}$$

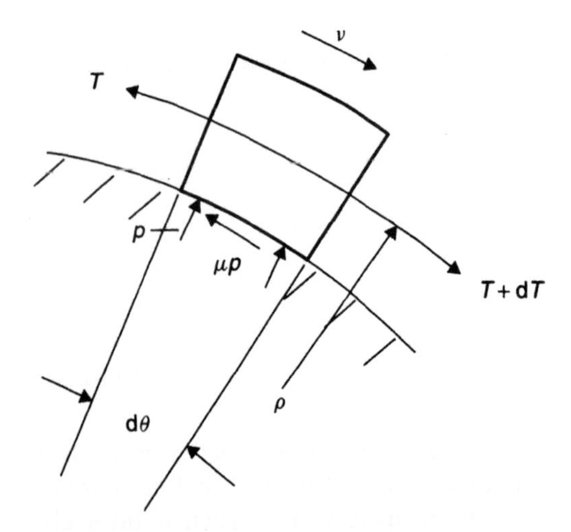

Figure 10.9 Element of sheet sliding on the surface between A and B in Figure 10.7.

or

$$T_B = T_A \exp\left(\mu\frac{\pi}{2}\right) \tag{10.24}$$

It is assumed that the tension is less than the yielding tension, so there is no change of thickness during this part of the process.

10.5.4 Unbending at B

During unbending at B there will be an increase in tension and a decrease in thickness as at A above and these can be determined from relations similar to Equations 10.20 and 10.21 using the tension T_B instead of T_0. The final tension in the process in Figure 10.7 is $T_f = T_B + \Delta T_B$.

10.5.5 (Worked example) drawing over a radius

Sheet is drawn over a radius under plane strain conditions as shown in Figure 10.7. The initial thickness is 2 mm and the tool radius is 8 mm (i.e. $\rho = 9$ mm). The material has a constant plane strain flow stress of 300 MPa. The back tension is 250 kN/m and the friction coefficient is 0.08. Assuming an efficiency factor of 0.8, determine by approximate methods the thickness of the sheet and the tension at exit.

Solution. The yield tension at the first bend is

$$T_y = 300 \times 10^6 \times 2 \times 10^{-3} = 600\,\text{kN / m}$$

From Equation 10.2, the increment in tension at the first bend is

$$\Delta T = \frac{300 \times 10^6 (2 \times 10^{-3})^2}{4 \times 0.8 \times 9 \times 10^{-3}} \left\{1 + \left(\frac{250}{600}\right)^2\right\} = 49\,\text{kN / m}$$

The change in thickness at the first bend is

$$\frac{\Delta t}{2} = \frac{2}{2 \times 9}\left(\frac{250}{600}\right) = 0.046\text{mm}$$

i.e.

$$\Delta t = 0.092\text{mm and } t = 1.91\text{mm}$$

The tension after the first bend is $250 + 49 = 299$ kN/m. The tension after the sheet has moved over the curved surface, from Equation 10.24, is

$$T = 299 \exp(0.08 \times \pi/2) = 339\,\text{kN / m}$$

The yield tension at the unbend is

$$T_y = 300 \times 10^6 \times 1.91 \times 10^{-3} = 573\,\text{kN / m}$$

The increase in tension as the sheet straightens is

$$\Delta T = \frac{300 \times 10^6 (1.91 \times 10^{-3})^2}{4 \times 0.8 \times 9 \times 10^{-3}} \left\{ 1 + \left(\frac{339}{573} \right)^2 \right\} = 51\,\text{kN / m}$$

and the exit tension is $299 + 51 = 350\,\text{kN/m}$.

The change in thickness at the unbend is

$$\frac{\Delta t}{t} = -\frac{1.91}{2 \times 9} \left(\frac{339}{573} \right) = -0.0628$$

and the thickness increment is $-0.0628 \times 1.91 = -0.12$, and the final thickness is $1.91 - 0.12 = 1.79$ mm.

10.6 Draw-beads

In draw die forming as in Chapter 4, draw-beads are used to generate tension in the sheet. A draw-bead is illustrated schematically in Figure 10.10. The sheet enters at the left with a small tension T_0, and then undergoes a series of bending or unbending processes marked by the broken lines in the diagram.

At each bending or unbending point, following Equation 10.20, the tension will increase, by

$$\Delta T = \frac{1}{4\eta} \frac{T_y}{(\rho/t)} \left\{ 1 + \left(\frac{T}{T_y} \right)^2 \right\}$$

and the thickness decrease, as indicated in Equation 10.21, by

$$\frac{\Delta t}{t} = -\frac{1}{2(\rho/t)} \left(\frac{T}{T_y} \right)$$

Between the points where the curvature changes, there will be an increase in tension, ΔT_f, due to friction, of

$$T + \Delta T_f = T \exp(\mu\theta)$$

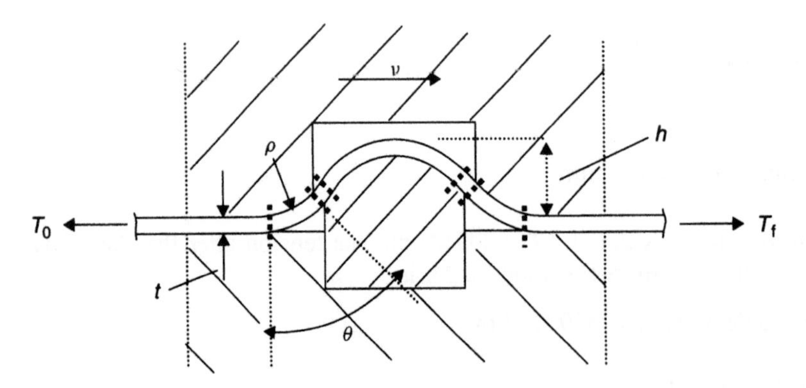

Figure 10.10 A draw-bead used to increase the tension in a sheet from T_0 to T_f as it passes through from left to right in a draw die.

where the angle of wrap θ is related to the depth of engagement h as shown in Figure 10.10.

The tension generated by a bead can be increased by reducing the bend ratio ρ/t, and increasing the depth of the bead h, which will increase the angle θ.

For a given bead, the tension will increase with the flow stress of the sheet S, with the incoming thickness t_0 and with the friction coefficient μ. It may be seen from Chapter 4 that in order to maintain a constant strain distribution in a part, the tension applied should increase if the above variables increase. Therefore a draw-bead is, to some extent, a self-compensating device in a draw die that will adjust the tension as material properties and friction change.

10.7 Exercises

Ex. 10.1 A sheet of aluminium, 1.85 mm, thick and having a constant plane strain flow stress of 180 MPa is curved over a form block having a radius of 600 mm. If the plane strain elastic modulus is 78 GPa, determine the tension required to (a) initiate plastic deformation, and (b) to make the sheet fully plastic.
[Ans: 111, 333 kN/m]

Ex. 10.2 For the operation in Exercise **10.1**, determine the tension and the moment when the plastic deformation zone has penetrated to the mid-surface. What is the final radius of curvature after unloading? Sketch the approximate form of the residual stress distribution in the sheet.
[Ans: 278 kN/m, 34 Nm/m, 1.19 m]

Ex. 10.3 Steel sheet is curved by stretching over a frictionless form block as in Figure 10.1. The radius of curvature of the block is 2.5 m. The final mean plastic strain in the sheet is 0.012 (1.2%). Aluminium sheet is substituted and the same final strain preserved. The plane strain elastic modulus and stress strain curves and the thickness are given below. Compare the final radius of curvature of each sheet.
[Ans: 2.56 v. 2.59 m]

Material	*E′ GPa*	*Stress, strain, MPa*	*Thickness, mm*
Steel	220	$\sigma_1 = 700\varepsilon_1^{0.2}$	0.8
Aluminium	78	$\sigma_1 = 400\varepsilon_1^{0.2}$	1.2

Ex. 10.4 A steel sheet is drawn over a radius $\rho = bt$, as in Figure 10.7. If the back tension is 60% of the yield tension in plane strain, show how the stress σ_1/S and the thickness reduction factor $-\Delta t/t_0$ vary with the bend ratio $b = \rho/t$ in the range $3 < b < 10$ for the first bend. Assume the efficiency factor is 1.

11 Hydroforming

11.1 Introduction

In hydroforming, or fluid pressure forming, sheet is formed against a die by fluid pressure. In many cases, a flexible diaphragm is placed on the sheet and it is then formed into a female die cavity as shown in Figure 11.1. The advantage of the process is that die construction is simpler and the process may be economical for making smaller numbers of parts. A disadvantage is that very high pressures may be required and the cycle time is greater than for stamping in a mechanical press.

Hydroforming is also used to form tubular parts such as brackets for bicycle frames or pipe fittings, as shown in Figure 11.2. Axial force may be applied to the tube as well as internal pressure; this creates compressive stress in one direction so that elements of the tube deform without thinning and tearing is delayed. With specially designed forming machines, a large number of parts can be formed by this process at low cost.

Another application is forming tubular parts such as vehicle frame components. A round tube is bent and then placed in a die as shown in Figure 11.3. It is then pressurized internally and formed to a square section.

In this chapter we do not attempt the analysis of a complete process, but break the process down into elements of simple geometry and model each of these separately. This will illustrate the limits of the process and show the influence of friction, geometry and material properties. The free expansion of a tube is first studied. Forming a square section from a round tube in the so-called *high-pressure* process is analysed and then it is shown how some of the process limits in this process can be overcome in a sequential forming process.

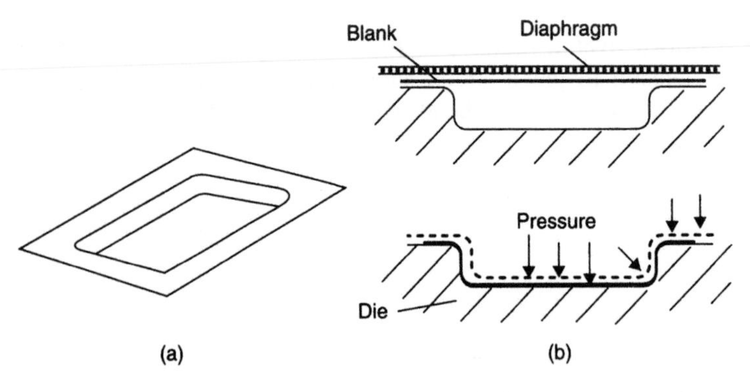

Figure 11.1 (a) A typical sheet metal part. (b) The arrangement for pressure, or hydroforming into a female die.

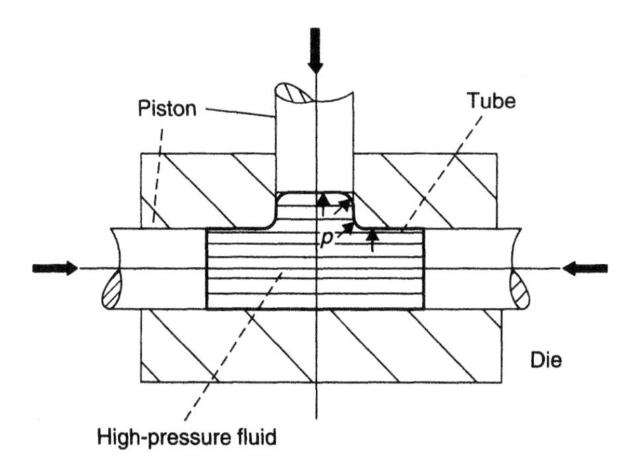

Figure 11.2 Hydroforming a tube component with pressure and axial force.

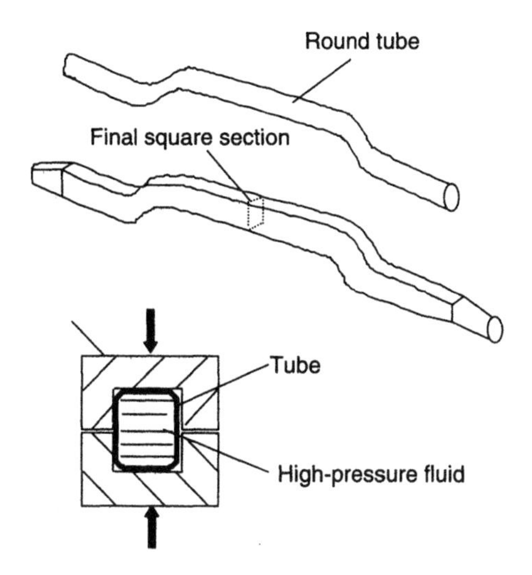

Figure 11.3 Method of hydroforming a bent square-section component from a round tube.

11.2 Free expansion of a cylinder by internal pressure

Expansion of a round tube without change in length is analysed. The tube will deform in plane strain, i.e. the strain in the axial direction will be zero. Initially the tube will remain circular and the radius will increase. The expansion of a cylindrical element in this mode is illustrated in Figure 11.4. The strain and stress states, for an isotropic material, are:

$$\varepsilon_\theta; \qquad \varepsilon_\phi = \beta\varepsilon_\theta = 0; \quad \varepsilon_t = -(1+\beta)\varepsilon_\theta = -\varepsilon_\theta;$$

$$\sigma_\theta; \qquad \sigma_\phi = \alpha\sigma_\theta = \frac{1}{2}\sigma_\theta; \quad \sigma_3 = 0 \tag{11.1}$$

i.e. $\beta = 0, \quad \alpha = \dfrac{1}{2}$

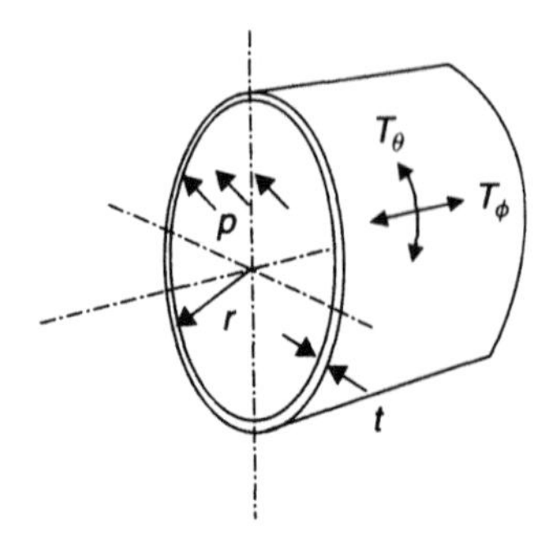

Figure 11.4 Element of a circular tube with internal pressure.

If the material properties obey the stress–strain law

$$\overline{\sigma} = K(\overline{\varepsilon})^n$$

When the material is deforming we obtain from Equations 2.18(b) and 2.19(c)

$$\sigma_\theta = \frac{2}{\sqrt{3}}\overline{\sigma} = \frac{2}{\sqrt{3}}\sigma_f \quad \text{and} \quad \overline{\varepsilon} = \frac{2}{\sqrt{3}}\varepsilon_1 \tag{11.2}$$

From Section 7.2, the principal radii at any region of the tube are

$$\rho_2 = \infty \quad \text{and} \quad \rho_1 = r \tag{11.3}$$

From Equation 7.3, the hoop tension is

$$T_\theta = \sigma_\theta t = pr \tag{11.4}$$

and as this is plane strain

$$T_\phi = \alpha T_\theta = \frac{1}{2}T_\theta$$

For the tube to yield, the pressure, from Equations 11.2 and 11.4, is

$$p = \frac{2}{\sqrt{3}}\frac{\overline{T}}{r} = \frac{2}{\sqrt{3}}\sigma_f\frac{t}{r} \tag{11.5}$$

If the tube is initially of thickness t_0 and radius r_0, the current hoop strain and thickness are given by

$$\varepsilon_\theta = \ln\frac{r}{r_0} \quad \text{and} \quad t = t_0 \exp\varepsilon_t = t_0\exp(-\varepsilon_\theta) = \frac{t_0 r_0}{r} \tag{11.6}$$

From this, we obtain the pressure characteristic for expansion as

$$p = \frac{2}{\sqrt{3}}K\left(\frac{2}{\sqrt{3}}\ln\frac{r}{r_0}\right)^n t_0\frac{r_0}{r^2} \tag{11.7}$$

From the form of Equation 11.7, we see that for a strain-hardening material, $n > 0$, the pressure will tend to increase as the material deforms. On the other hand, if the tubular element is allowed to expand freely, the tube wall will thin and the radius increase; both effects will tend to decrease the pressure. At some point, the pressure will reach a maximum as the opposing effects balance. Differentiating Equation 11.7 shows that the maximum pressure is when $\varepsilon_\theta = n/2$. Beyond this, the tube would swell locally.

We need also to consider the possibility of the tube wall necking and splitting within the locally bulged region. As this is plane strain deformation, the loading path will be along the vertical axis in the strain space, Figure 5.16. Splitting would be expected approximately when the hoop strain has approximately the value n. Thus the limiting case is when

$$\ln \frac{r^*}{r_0} = n \quad \text{and} \quad p^* = \frac{2}{\sqrt{3}} K \left(\frac{2}{\sqrt{3}} n \right)^n \frac{t_0}{r_0} \exp(-2n) \tag{11.8}$$

where r^* is the radius at which the tube splits. It should be noted that tube material is likely to be anisotropic. If properties are measured from tensile tests in the axial direction, the strain hardening index in the circumferential direction may not be known accurately and splitting at strains considerably less than predicted by Equation 11.8 may occur.

11.3 Forming a cylinder to a square section

A common operation is forming a round tube into a square die as illustrated in Figure 11.3; in the middle of the part, the tube will deform in plane strain. Various stages of the process are shown in Figure 11.5.

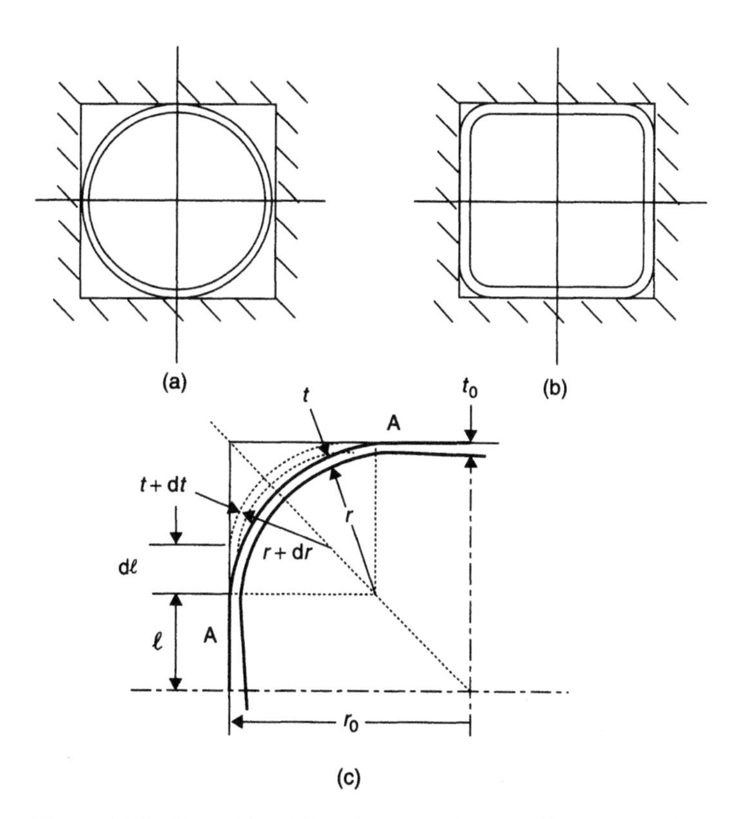

Figure 11.5 Expanding (a) a circular tube into (b) a square die by fluid pressure; (c) shows an increment in the process at the corner.

In Figure 11.5(c), the tube has been partially expanded so that the wall is touching the die up to the point A. The contact length is l and assuming that the thickness is small compared with the radius, the current contact length is $l = r_0 - r$. During an increment in the process, the contact length increases to $l + \mathrm{d}l$ and the corner radius decreases to $r + \mathrm{d}r$, where $\mathrm{d}r$ is a negative quantity, i.e. $\mathrm{d}r = -\mathrm{d}l$.

In the contact zone, the tube will be pressed against the die wall by the internal pressure. If the material slides along the die, friction will oppose the motion and the tension will vary around the tube wall. At some point, the tension will be insufficient to stretch the wall and there will be a sticking zone as shown in Figure 11.6.

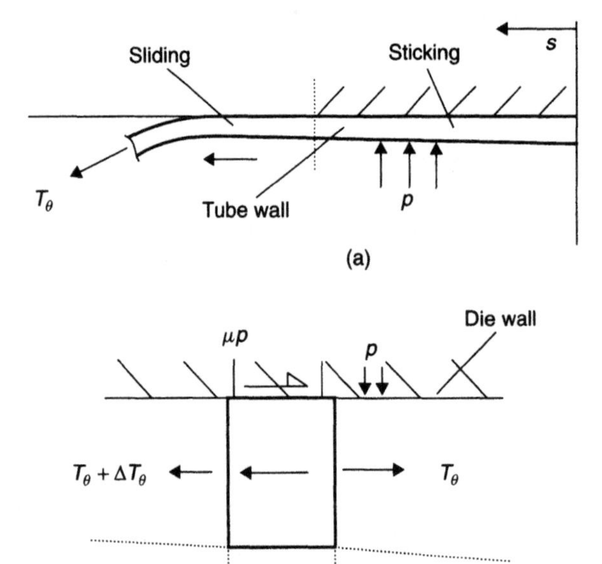

Figure 11.6 (a) Part of the tube wall in contact with the die during forming of a cylinder to a square section. (b) Element of the tube wall in the sliding zone.

The equilibrium equation for the element in Figure 11.6(b) is

$$T_\theta + \mathrm{d}T_\theta = T_\theta + \mu p\,\mathrm{d}s\,1$$

or

$$\frac{\mathrm{d}T_\theta}{\mathrm{d}s} = \mu p \tag{11.9}$$

The deformation process is stable as long as the tension increases with strain so we assume here that the tension in the unsupported corner will continue to increase as the radius becomes smaller. Referring to Section 3.7, the slope of the tension versus strain curve for plane strain is positive for strains $\varepsilon_\theta \le n$. Thus in Figure 11.6, the tension to yield the tube will increase as the tube wall thins. For the tube in contact with the die, the greatest tension will be at the tangent point. Equation 11.9 shows that due to friction, the

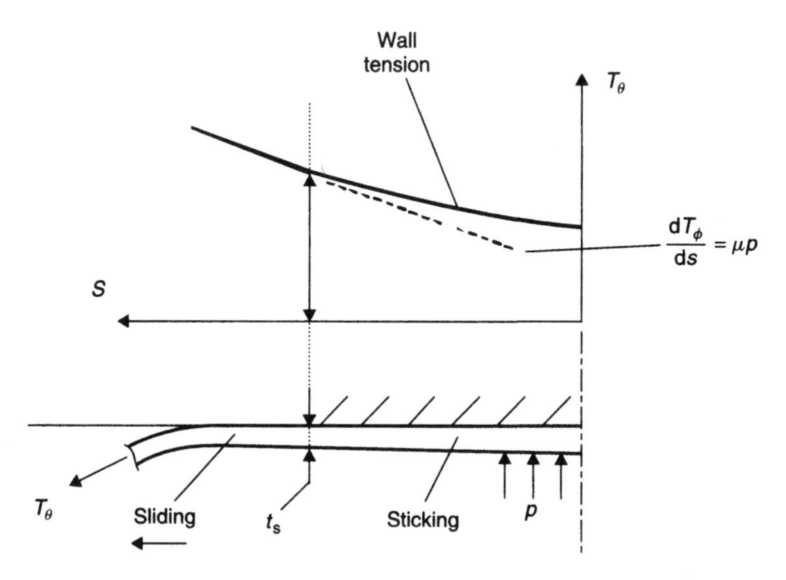

Figure 11.7 Distribution of tension in that part of the tube in contact with the die wall during forming of a round tube to a square section.

tension decreases linearly towards the centre-line; the distribution is shown in Figure 11.7. To the right of the point where sliding ceases, the tension in the tube wall is less than that required for yielding and there is no further deformation in this sticking region. The critical point is where the thickness is t_s. For a material obeying the stress–strain law

$$\overline{\sigma} = K(\overline{\varepsilon})^n$$

from Equations 11.2, the hoop stress is given by

$$\sigma_\theta = \frac{2}{\sqrt{3}} K \left(\frac{2}{\sqrt{3}} \varepsilon_\theta \right)^n$$

As this is a plane strain process,

$$\varepsilon_\theta = -\varepsilon_t = \ln \frac{t_0}{t}$$

The tension at the critical point in Figure 11.7 separating sliding and sticking is

$$T_{\theta_s} = \sigma_\theta t_s = \frac{2}{\sqrt{3}} K \left(\frac{2}{\sqrt{3}} \ln \frac{t_0}{t_s} \right)^n t_s \tag{11.10}$$

In the sticking zone, there is no sliding and the slope of the tension curve in Figure 11.7 will be less than μp.

The distribution of thickness in the wall can be determined by an incremental analysis. In this work, the extreme cases, either with no friction at the die wall or with sticking friction along the entire contact length will be considered.

11.3.1 Tube forming in a frictionless die

If contact between the tube and the die is frictionless, at any instant the tension and also the thickness at any point around the tube will be uniform. The current perimeter length of one-quarter of the tube in Figure 11.5 is

$$\frac{\pi}{2}r + 2l = \frac{\pi}{2}r + 2(r_0 - r) = 2r_0 - r\left(2 - \frac{\pi}{2}\right) \tag{11.11}$$

As there is no change in volume of the tube material

$$t\left[2r_0 - r\left(2 - \frac{\pi}{2}\right)\right]1 = t_0\frac{\pi}{2}r_0 1$$

or

$$t = \frac{t_0}{\left\{\frac{4}{\pi} - \frac{r}{r_0}\left(\frac{4}{\pi} - 1\right)\right\}} \tag{11.12}$$

As this is a plane strain process, the hoop strain is

$$\varepsilon_\theta = -\varepsilon_t = \ln\frac{t_0}{t}$$

and for a material obeying the stress–strain law $\overline{\sigma} = K\overline{\varepsilon}^n$, the hoop stress is

$$\sigma_\theta = \frac{2}{\sqrt{3}}K\left(\frac{2}{\sqrt{3}}\ln\frac{t_0}{t}\right)^n \tag{11.13}$$

The pressure required to deform the unsupported corner of radius r can be determined from Equation 11.4.

11.3.2 Tube forming with sticking friction (or very high friction)

If the tube sticks to the die wall as soon as it touches it, then for a circular tube that initially just fits inside the die as shown in Figure 11.5(a), the thickness at the first point of contact will be t_0. As the tube becomes progressively attached to the wall, the thickness at the tangent point will decrease so that at A, in Figure 11.5(c), it has the value t. For a unit length perpendicular to the plane of the diagram, the volume of material in the arc AA is $\pi rt/2$. This volume will remain unchanged during the increment, and as this is plane strain, equating the volumes before and after the increment, we obtain

$$\left[\frac{\pi}{2}(r + \mathrm{d}r) + 2\mathrm{d}l\right](t + \mathrm{d}t) = \frac{\pi}{2}rt$$

From above, $l = r_0 - r$, and $\mathrm{d}l = -\mathrm{d}r$, hence

$$\frac{\mathrm{d}t}{t} = \left(\frac{4}{\pi} - 1\right)\frac{\mathrm{d}r}{r} \tag{11.14}$$

Integrating with the initial conditions $t = t_0$ at $r = r_0$, we obtain

$$t = t_0\left(\frac{r}{r_0}\right)^{\frac{4}{\pi}-1} \tag{11.15}$$

The hoop strain and pressure to continue the deformation can then be determined following the same approach as in Section 11.3.1. The wall thickness will be non-uniform varying from the initial thickness at the centre-line of the die to a minimum at the unsupported corner radius. For a given corner radius, the corner thickness will be less than for the frictionless case and failure could occur earlier in the process.

11.3.3 Failure in forming a square section

The above analysis neglects the effect of unbending under tension at the tangent point A in Figure 11.5. This will be similar to the effects described in Section 10.5, and cause additional thinning and work-hardening of the tube wall. In plane strain forming of a tube of uniform section, the process is possible provided that the tension in the deforming wall continues to increase with forming. If it reaches a maximum, necking and failure of the tube will take place. As indicated in Section 5.4, for plane strain, the limit for a power law strain-hardening material is approximately when $\varepsilon_\theta = n$.

In forming the corner radius, the pressure required will increase as the radius decreases, as shown by Equation 11.5. Even though very high pressure equipment may be used, the limitation on corner radius is usually significant and this, together with the forming limit strain, are the first things that should be calculated in preliminary design.

11.4 Constant thickness forming

In plane strain expansion, as indicated above, the material will split approximately when the hoop strain reaches the value of the strain-hardening index n. The strain path is illustrated in Figure 11.8(a). To obtain the required strain in the part, high strain-hardening material is used, but this exacerbates the problem of high pressure for forming.

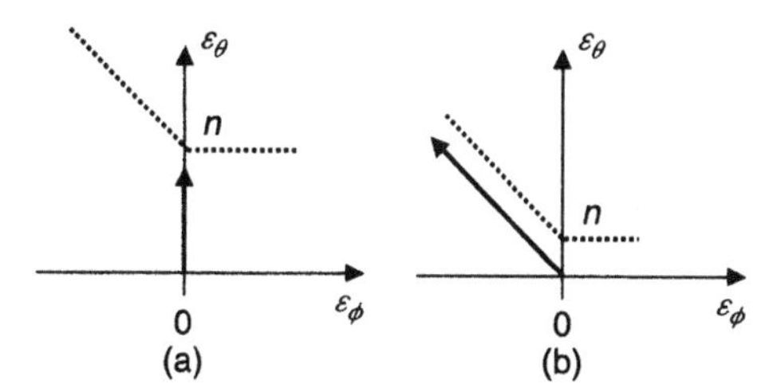

Figure 11.8 Strain paths for (a) plane strain and (b) constant thickness forming.

Large strains are potentially possible in processes such as those illustrated in Figure 11.2, and it would be advantageous to choose a low strain-hardening material. To avoid splitting, a constant thickness strain path $\beta = -1$ would be the objective and this strain path together with the forming limit for a low n material is illustrated in Figure 11.8(b). The design of processes that achieve constant thickness deformation is not easy, but we consider some cases of simple geometry.

11.4.1 Constant thickness deformation for a tube expanded by internal pressure

In a constant thickness process, the hoop and axial tensions and stresses would be equal and opposite. The hoop tension, from Equation 11.4, is $T_\theta = pr$; the axial tension or traction,

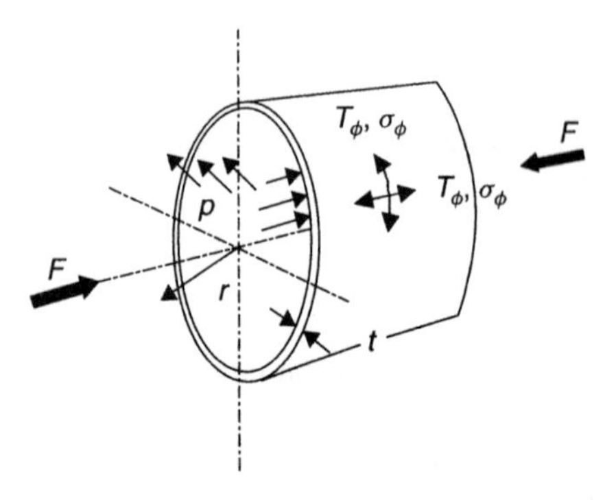

Figure 11.9 Cylindrical element deforming under constant thickness conditions with internal pressure and axial compression.

which is compressive, is $T_\phi = -pr$. To achieve this, an axial force must be applied to the tube as illustrated in Figure 11.9; the axial force is

$$F = -2\pi r T_\phi = 2\pi r p r = 2\pi r^2 p \tag{11.16}$$

From Equations 2.18 and 2.19, the effective strain and stress are

$$\bar{\varepsilon} = \frac{2}{\sqrt{3}}\varepsilon_\theta = \frac{2}{\sqrt{3}}\ln\frac{r}{r_0} \quad \text{and} \quad \sigma_\theta = \frac{1}{\sqrt{3}}\bar{\sigma} \tag{11.17}$$

where σ_θ is given by Equation 11.4. For a given expansion in a material obeying the stress–strain law $\bar{\sigma} = K(\bar{\varepsilon})^n$, and noting that the thickness remains constant, we obtain the pressure required to continue the process as

$$p = \sigma_\theta\frac{t}{r} = \frac{1}{\sqrt{3}}K\left(\frac{2}{\sqrt{3}}\ln\frac{r}{r_0}\right)^n\frac{t_0}{r} \tag{11.18}$$

Comparing this with Equation 11.7, we see that the pressure is reduced for constant thickness expansion, compared with plane strain, but of course an axial force is required and this will do work as well as the pressure.

11.4.2 Effect of friction on axial compression

To achieve axial compression on an element, a force is applied to the end of the tube as shown in Figure 11.2. The effect of this force is local because friction between the tube and the die will cause it to diminish with distance from the point of application of the force. The case for a simple tube is shown in Figure 11.10.

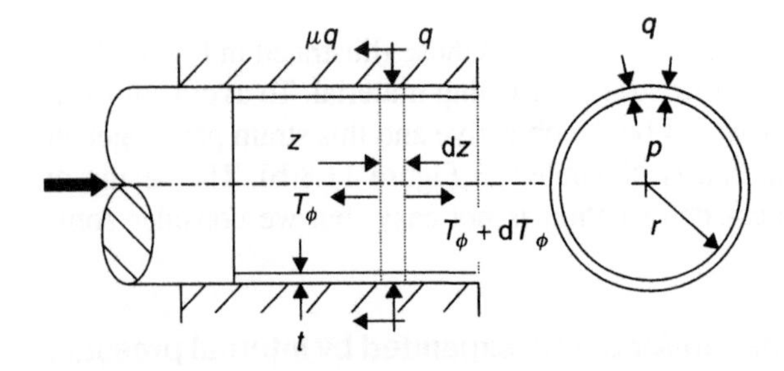

Figure 11.10 Effect of friction on the axial compression of a tube.

In this diagram, a plunger on the left applies compression to a tube. At some distance z the tension on one side of an element of width $\mathrm{d}z$ is T_ϕ, and on the other side, $T_\phi + \mathrm{d}T_\phi$. Around this elemental ring there is a contact pressure q and a friction stress μq. The equilibrium equation for the element is

$$2\pi r(T_\phi + \mathrm{d}T_\phi) = 2\pi r T_\phi + 2\pi r \mu q \, \mathrm{d}z$$

or

$$\mathrm{d}T_\phi = \mu q \, \mathrm{d}z \tag{11.19}$$

This shows that the tension, or traction increases, i.e. becomes more tensile as z increases. It starts as most compressive at the plunger and the compression decreases linearly with distance from the end of the tube. For this reason it is not possible to obtain axial compression in the middle of parts such as shown in Figure 11.3, but in shorter parts, such as Figure 11.2, axial compression can be effective in preventing thinning and necking. In parts of complicated shape, such as in Figure 11.2, it is found that to ensure that thinning does not occur in any critical regions, some places will become thicker, but this is usually acceptable.

11.5 Low-pressure or sequential hydroforming

In forming parts such as in Figure 11.3, a technique has evolved that avoids the use of very high pressures. To form a square section, an oval tube is compressed during closure of the die and the internal fluid only serves to keep the tube against the die as shown in Figure 11.11.

In this process, the periphery of the tube remains approximately constant and tearing is not likely to be a limiting factor. The mechanics of the process can be illustrated by considering the case of a circular arc of a tube being formed into a corner as shown in Figure 11.12. In this case the process is symmetric about the diagonal; this is not quite the same as that shown in Figure 11.11, but will illustrate the principle.

In forming the tube into the corner, the wall is bent to a sharper radius at B and straightened at A. The bending moment diagram is shown in Figure 11.12(b). The stress distribution at B is shown in Figure 11.13, for a rigid, perfectly plastic material.

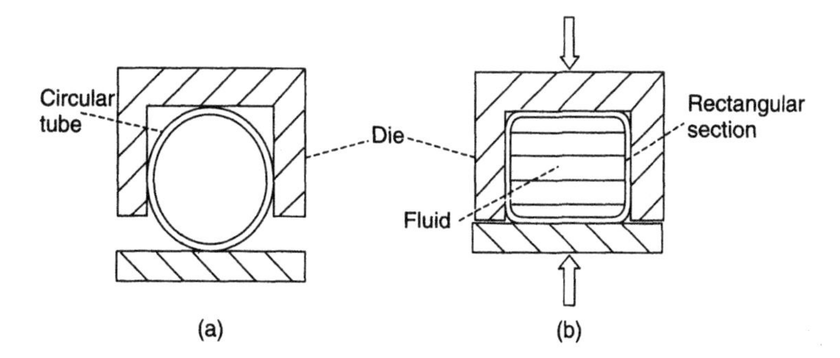

Figure 11.11 Forming (a) an oval tube into (b) a square or rectangular section in a low-pressure hydroforming process.

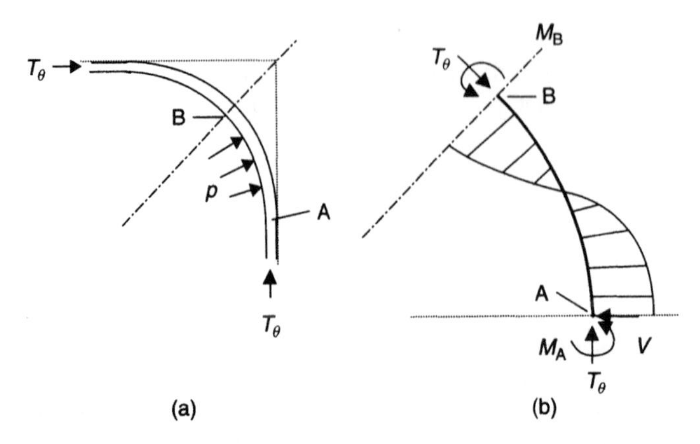

Figure 11.12 Deformation of the corner of a tube in a low-pressure hydroforming process. (a) The forces acting, and (b), the bending moment diagram on the arc AB.

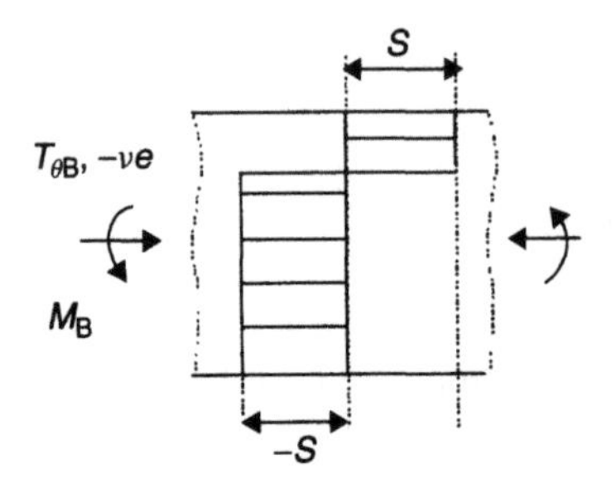

Figure 11.13 Stress distribution at the corner of a tube being deformed as in Figure 11.12.

As shown in Section 10.4, the bending moment required to deform the tube at B is

$$M_B = \frac{St^2}{4}\left[1 - \left(\frac{T_{\theta B}}{T_y}\right)^2\right] = M_p\left[1 - \left(\frac{T_{\theta B}}{T_y}\right)^2\right] \tag{11.20}$$

where S is the plane strain flow stress, T_y the yield tension, and M_p the moment to bend the wall in the absence of tension or compression, i.e. the fully plastic moment determined in Section 6.5.2. Thus by applying compression to the tube wall, the tube can be bent easily at the corner and the tangent point; a plastic hinge will form at B to bend the tube, and one at A to straighten it. In this simple analysis, bending will only occur at A and B and the region between will remain unchanged. This is not very realistic as material is never completely rigid, perfectly plastic, but it is observed that in this process, the radius of curvature in the unsupported corner is not constant; it is least at the point B.

In performing this operation, the tube is filled with fluid, usually water, and sealed as it is placed in the die as in Figure 11.11. As the internal volume diminishes in forming the oval to the square section, fluid is expelled from the tube. The pressure is regulated using a pressure control valve and is maintained at a sufficient level to keep the tube against the die walls and prevent wrinkling. Once the die is closed, the pressure may be increased to improve the shape. It may be seen that the overall perimeter of the tube does not change very much during forming and therefore splitting is avoided.

11.6 Summary

In this chapter, several example of forming using fluid pressure have been examined. There are many other applications of this and the following factors should be borne in mind in considering fluid forming.

- Very high fluid pressures are required to form small fillet radii.
- Forming equipment becomes expensive as the pressures become high.
- Very high strains can be achieved if compressive forces are applied to the part as well as pressure and approximately constant thickness deformation obtained.
- Forming pressures can be reduced if controlled buckling under compressive forces is obtained.

11.7 Exercises

Ex. 11.1 A circular tube with a radius R and thickness t_0 is deformed into a die with a square cross-section through high pressure hydroforming. Find the relation between the corner radius and the internal pressure in the two cases:

(a) There is no friction at the die metal interface.
(b) The tube material fully sticks to the die surface.

The material obeys effective stress–strain law $\bar{\sigma} = K(\varepsilon_o + \bar{\varepsilon})^n$.

Ex. 11.2 A mild steel tube of 180 mm diameter and thickness 4 mm is to be expanded by internal pressure into a square section. The maximum pressure available is 64 MPa. The material has a stress–strain curve

$$\bar{\sigma} = 700(\bar{\varepsilon})^{0.2}\ \text{MPa}$$

Determine the minimum corner radius that can be achieved if the die is functionless.
[Ans: 31 mm]

Ex. 11.3 In a tube hydroforming process, a square section was formed from a circular tube. The initial tube had a radius of R. The current corner radius is r. The maximum thickness along the tube was t_1. The corner thickness was t_2. The material strain-hardens according to the relation $\bar{\sigma} = K(\varepsilon_o + \bar{\varepsilon})^n$.

Assuming that the thickness varies linearly along the tube wall, calculate the average friction coefficient between the tube wall and die surface. Assume the tube just touches the die wall at the start of the process and that there is no sticking region.

Ex. 11.4. Find the pressure required to expand a circular tube from initial radius r_0 to a final radius r. The material hardens as $\sigma = K\varepsilon^n$, for two cases

(a) Tube ends are free to move axially.
(b) Tube ends are restricted from axial movements.

Calculate also the strain at maximum pressure.
[Ans: 2n/3, n/2]

Appendix A1
Yielding in three-dimensional stress state

A.1.1 Introduction

In Chapter 1 the conditions for yielding of a material element were given for plane stress, when one principal stress is zero. There are cases in sheet forming where the stress normal to the surface is not negligible and the principal stresses are as shown in Figure A1.1. The conditions for yielding when the current flow stress of the element is σ_f are given below.

A1.2 Yield criteria

A1.2.1 Tresca yield criterion

If the principal stresses are ranked in order of magnitude, i.e. the most tensile first and the most compressive last, the Tresca condition is as given in Equation 2.10:

$$|\sigma_{max.} - \sigma_{min.}| = \sigma_f \qquad \text{(A1.1)}$$

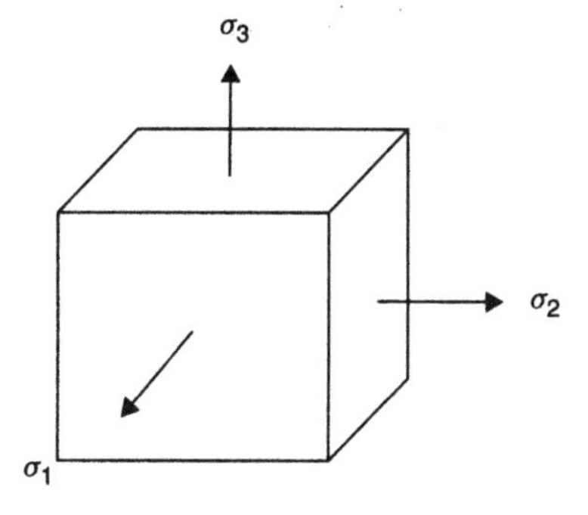

Figure A1.1 Principal stresses on an element in a three-dimensional stress state.

A1.2.2 von Mises yield criterion

The condition that yielding occurs when the root-mean-square value of the maximum shear stresses reaches a critical condition was given in Section 2.4.2. This can be written for the general case on the left, and for the tensile test on the right as

$$\sqrt{\frac{\tau_1^2+\tau_2^2+\tau_3^2}{3}}=\sqrt{\frac{2\left(\sigma_f/2\right)^2}{3}}$$

Substituting for the maximum shear stresses shown in Figure 2.4 and simplifying, we obtain the von Mises yielding condition as

$$\sqrt{\frac{1}{2}\left[(\sigma_1-\sigma_2)^2+(\sigma_2-\sigma_3)^2+(\sigma_3-\sigma_1)^2\right]}=\sigma_f \tag{A1.2}$$

As discussed in Section 2.4, the hydrostatic component of stress should not influence yielding and therefore it follows that the yielding criterion can also be written in terms of the reduced or deviatoric stresses, i.e.

$$\sqrt{\frac{3}{2}\left[\sigma_1'^2+\sigma_2'^2+\sigma_3'^2\right]^2}=\sigma_f \tag{A1.3}$$

A1.3 Graphical representation of yield theories

In Figure A1.2. the stress state for an element that is yielding is represented by the *stress vector* 0P in the three-dimensional stress space. A stress state in which $\sigma_1=\sigma_2=\sigma_3$ would be on the line 0H which is known as the hydrostatic axis; this is equally inclined to the three principal stress axes. By considering the right-angled triangle 0NP it may be shown that the von Mises hypothesis is equivalent to the statement that an element will yield when the stress point P is a critical distance from the hydrostatic axis, i.e. when NP reaches a critical value. This value may be determined from the known value of NP when P represents the uniaxial stress state, i.e. when $\sigma_1=\sigma_f$, $\sigma_2=\sigma_3=0$.

The von Mises criterion may therefore be represented in the three-dimensional stress space as a cylinder whose axis coincides with the hydrostatic stress axis and passes through the points, $\pm\sigma_f$, on the principal stress axes. If the stress point P is inside the cylinder, the element is in an elastic state and does not yield. The region outside the cylinder represents stress states that the element cannot support. In strain-hardening materials, the radius of this cylinder will increase as the element strains, and the diagram only represents the yield condition at some instant in the process.

The flow rule given in Section 2.5 may be used to show that the strain vector representing the principal strain increments will be normal to this cylinder during some increment in the process.

In both these diagrams, the yield locus in plane stress, $\sigma_3=0$, is represented by the horizontal section through the cylinder or hexagon through the origin. This is the same as the loci shown in Figures 2.7 and 2.8.

It may be shown that the Tresca yield criterion may be represented by a hexagonal surface lying within the von Mises cylinder as shown Figure A1.3. The vertices of this hexagon will just touch the cylinder.

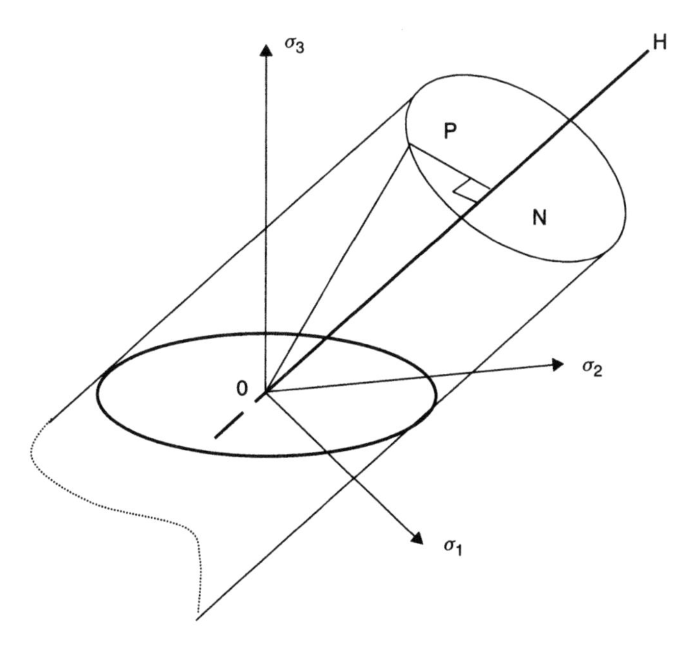

Figure A1.2 Representation of the von Mises yielding criterion as a cylinder in the principal stress space.

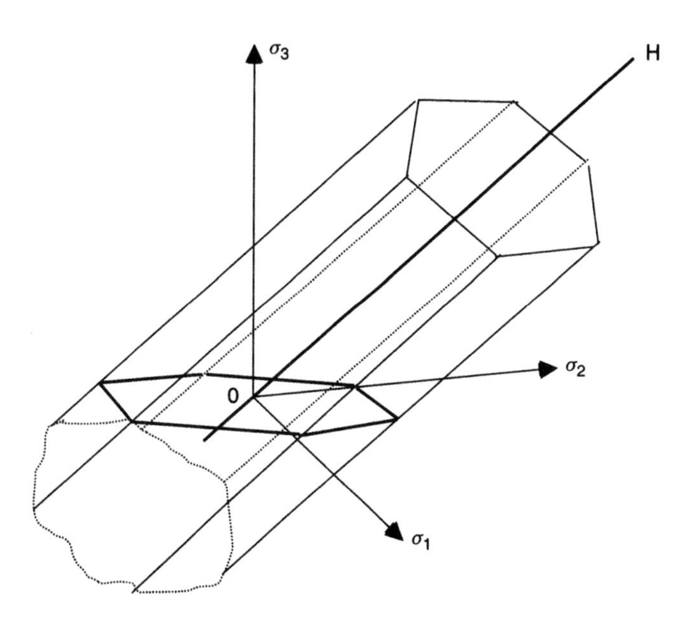

Figure A1.3 Representation of the Tresca yield criterion in the principal stress space.

Appendix 2

Large strains: an alternative definition

A2.1 Introduction

In Chapter 1 it was shown that engineering strains are unsuitable for modelling processes and incremental and natural strains were introduced. A problem exists in decomposing natural strains into Cartesian components and an alternative strain definition which may be more convenient for experimental strain analysis is introduced here.

We consider a unit principal cube as shown in Figure A2.1(a); after some large deformation, this would remain rectangular and of side

$$1 + E_1; \quad 1 + E_2; \quad \text{and} \quad 1 + E_3 \tag{A2.1}$$

where E_i can be considered as *large principal strains*. These are related to the natural principal strains by,

$$\varepsilon_i = \ln(1 + E_i) \tag{A2.2}$$

A2.2 Deformation of an element in terms of the large strains E

In sheet metal forming, we are usually interested in the membrane strains in the sheet, i.e. those in the plane of the sheet. For this reason, we examine here the deformation of an element in the sheet and assume that the principal direction, 3, is perpendicular to the sheet

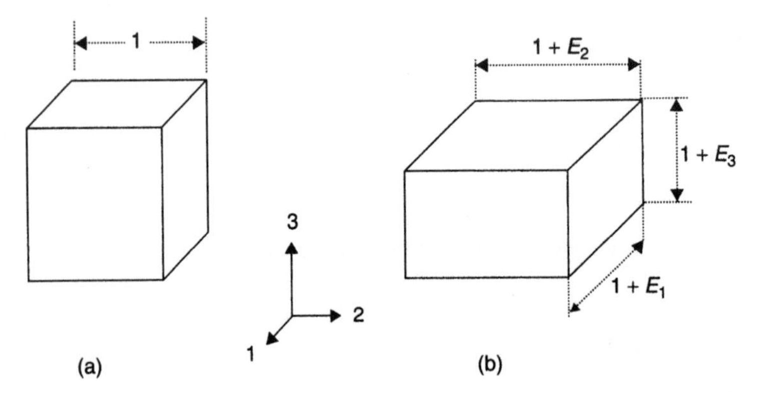

Figure A2.1 A unit principal cube (a) before deformation and (b) after deformation.

surface. In the 1,2, plane we consider a point P having the coordinates X and Y, in the principal directions 1 and 2. If the axes are fixed in the material, the relative displacement component of the point P are ΔX and ΔY as shown in Figure A2.2, where

$$\Delta X = E_1 X \quad \text{and} \quad \Delta Y = E_2 Y \tag{A2.3}$$

We now consider this displacement in non-principal axes $0x$ and $0y$ as shown in Figure A2.3.

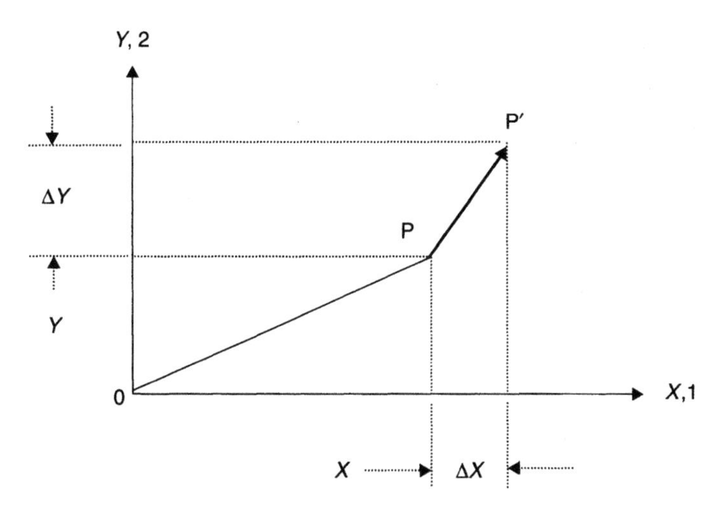

Figure A2.2 Displacement of a general point P in the 1,2 plane.

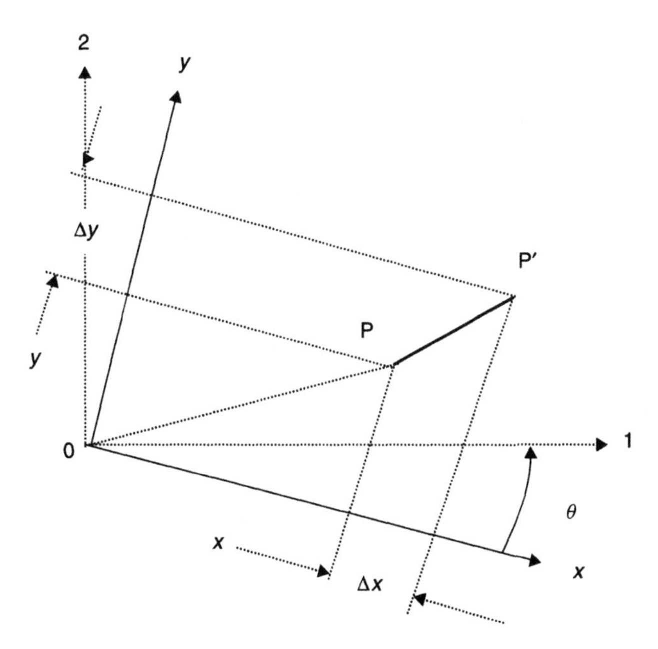

Figure A2.3 Displacement of a general point in non-principal axes.

The displacement is given by

$$\begin{pmatrix} \Delta x \\ \Delta y \end{pmatrix} = \begin{vmatrix} \cos\theta, -\sin\theta \\ \sin\theta, \cos\theta \end{vmatrix} \begin{pmatrix} \Delta X \\ \Delta Y \end{pmatrix} \tag{A2.4}$$

and the coordinates in the principal frame are given in terms of the non-principal frame by

$$\begin{pmatrix} X \\ Y \end{pmatrix} = \begin{vmatrix} \cos\theta . \sin\theta \\ -\sin\theta, \cos\theta \end{vmatrix} \begin{pmatrix} x \\ y \end{pmatrix} \tag{A2.5}$$

We now consider a non-principal unit line pair as shown in Figure A2.4.

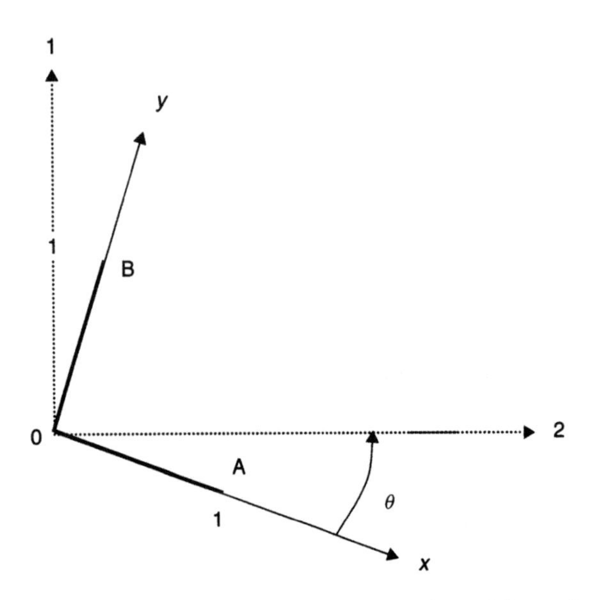

Figure A2.4 A unit, orthogonal, non-principal line pair in the 1,2 plane.

The coordinates of point A are (1, 0) and of B, (0, 1) and the displacement components from the above equations

$$\begin{aligned} \Delta x_a &= E_1 \cos^2\theta + E_2 \sin^2\theta = E_{xx} \\ \Delta y_a &= \frac{E_1 - E_2}{2} \sin 2\theta = E_{xy} = \Delta x_b = E_{yx} \\ \Delta y_b &= E_1 \sin^2\theta + E_2 \cos^2\theta = E_{yy} \end{aligned} \tag{A2.6}$$

The above equations define the Cartesian components of the large strains and these are illustrated in Figure A2.5. It may be seen that the lineal strain components E_{xx}, E_{yy} are the *projections* of the displacements on the original grid axes. The shear components E_{xy}, E_{yx} are the *offsets* from these axes. It will also be seen that each line rotates with respect

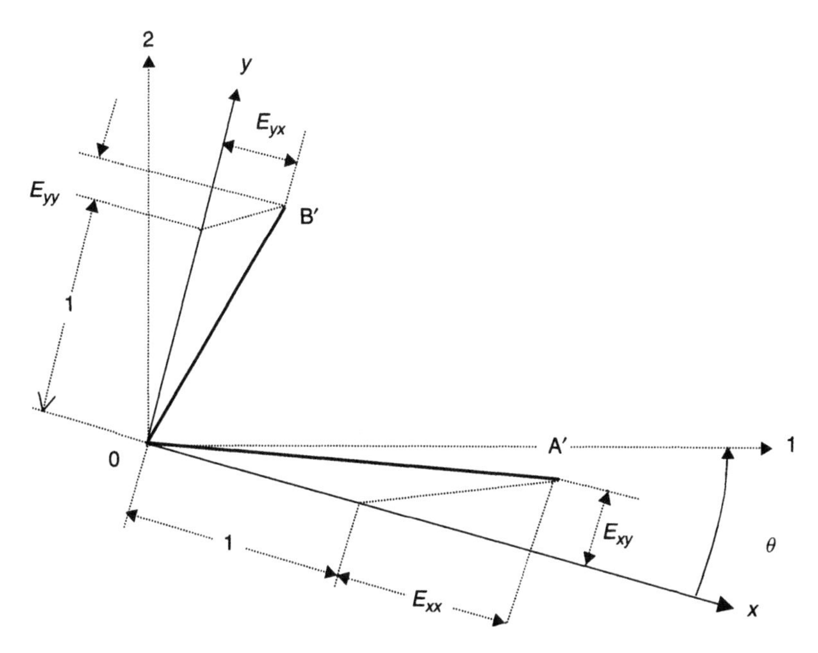

Figure A2.5 Deformation of a unit, orthogonal, non-principal line pair, illustrating the Cartesian components of large strain.

to the principal axes and the line pair is no longer orthogonal. The diagram also shows that as the shear components are equal the offsets of points A′ and B′ will be equal.

The Equations A2.6. are similar to those for the familiar small engineering strains; they have similar transformation equations and may be represented in a Mohr circle of strain. The principal strains are

$$E_{1,2} = \frac{E_{xx} + E_{yy}}{2} \pm \sqrt{\left(\frac{E_{xx} - E_{yy}}{2}\right)^2 + E_{xy}^2} \tag{A2.7}$$

and

$$\tan 2\theta = \frac{2E_{xy}}{E_{xx} - E_{yy}} \tag{A2.8}$$

A2.3 Experimental strain analysis

Measurement of strains from circles marked on the sheet has been mentioned previously. In some cases it is preferable to mark a sheet with an orthogonal grid of equi-spaced lines. If the intersections of these lines are measured, the strain distribution over the part can be determined. We illustrate this here for a simple line pair as shown in Figure A2.4. In the undeformed state, this is as shown in Figure A2.6(a); after deformation, the grid becomes a curvilinear mesh as shown in Figure A2.6(b). We assume the grid points are measured

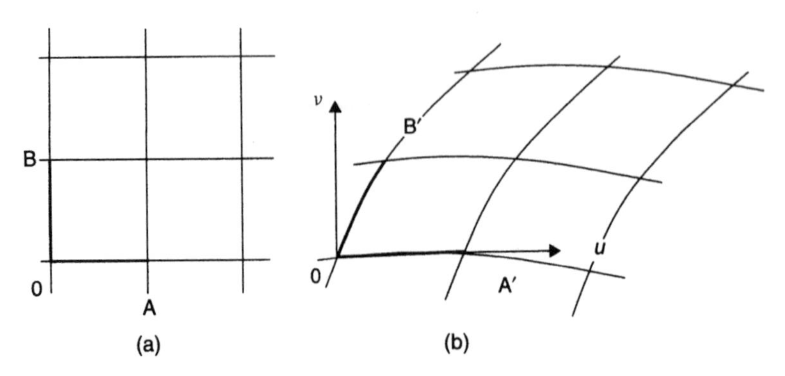

Figure A2.6 A non-principal, orthogonal line pair in (a) the undeformed state and (b) the deformed state.

in the local reference frame, $0u$, $0v$, as shown, where $0u$ passes through the point A′ and the axes lie in the plane A′0 B′.

The coordinates of the points in the measuring axes are

$$u_a, v_a; \quad \text{and} \quad u_b, v_b, \text{ where } v_a = 0$$

It may be shown that the following components exist:

$$\begin{aligned} b_{11} &= u_a^2 \\ b_{12} &= b_{21} = u_b \cdot u_a \\ b_{22} &= u_b^2 + v_b^2 \end{aligned} \tag{A2.9}$$

The principal large strain values are

$$(1 + E_{1,2})^2 = \left(\frac{b_{11} + b_{22}}{2}\right) \pm \sqrt{\left[\frac{(b_{11} - b_{22})}{2}\right]^2 + (b_{12})^2} \tag{A2.10}$$

The orientation of the principal axes *with respect to the original grid axes, $0x$ and $0y$,* is

$$2\theta = \tan^{-1}\frac{2b_{12}}{(b_{11} - b_{22})} + n\frac{\pi}{2} \tag{A2.11}$$

The argument of $\tan^{-1}$ is in the range $-\frac{\pi}{2}$ to $\frac{\pi}{2}$, and

$$\begin{aligned} n &= 0 \quad \text{if } b_{11} \geq b_{22}; \\ n &= 1 \quad \text{if } b_{11} < b_{22} \quad \text{and } b_{12} \geq 0; \\ n &= -1 \quad \text{if } b_{11} < b_{22} \quad \text{and } b_{12} < 0 \end{aligned}$$

In experimental strain analysis, it is often useful to show on the *undeformed* grid, the principal strains and their orientation as shown in Figure A2.7.

A2.4 Limitations of large strains

The large strains E defined in this appendix are useful in practical applications where familiarity with natural strains may not exist. The large strains are equivalent to elongations which is a more familiar concept. They also have the advantage that the Cartesian

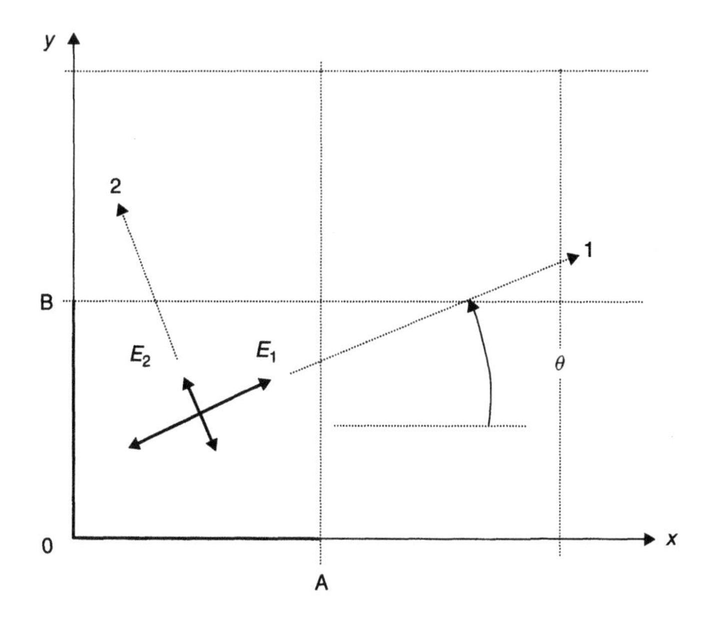

Figure A2.7 Principal strains illustrated in the undeformed axes for the line pair, 0A, 0B.

components form a symmetric tensor and that the Mohr circle of strain can be applied. A disadvantage is that in a *proportional* process, as defined in Chapter 2, the ratio of the principal large strains will not remain constant and proportional paths in large strain space will not be linear.

A2.5 Exercise

Ex. A2.5.1 (a) An orthogonal grid of 5 mm pitch is marked on a sheet. A line pair is measured in the axes, $0u$, $0v$, as illustrated in Figure A2.6(b). The coordinates are:

A'	6.017,	0
B'	1.741,	4.591

measured in mm. Determine the large principal strains, the principal natural strains and the orientation of the principal directions in the undeformed grid. Illustrate the results in a diagram similar to Figure A2.7.
(0.30,- 0.15, or 30%,- 15%) (0.262, -0.163) (30°)

(b) Determine the Cartesian components of large strain, E, in the axes, $0x$, $0y$. Illustrate these in a diagram similar to Figure A2.5.
(0.182, 0.048, 0.168)

Solutions

Ex. A2.5.1(a) In Figure A2.8, the line pair for a unit square in the undeformed state in the axes, Ox and Oy, is denoted by Oa and Ob. In the deformed state, this line pair becomes Oa' and Ob'. The coordinates in the axes Ou and Ov in the deformed state are:

$$u_a = 6.017/5 = 1.203 \quad v_a = 0$$

$$u_b = 1.741/5 = 0.348 \quad v_b = 4.591/5 = 0.918$$

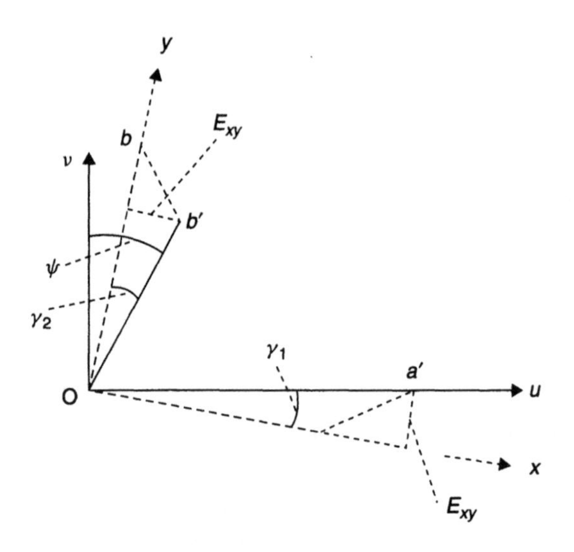

Figure A2.8

Solution A

Using Equations A2.9,

$$b_{11} = 1.203^2 = 1.447$$

$$b_{12} = 0.348 \times 1.203 = 0.419$$

$$b_{22} = 0.348^2 + 0.918^2 = 0.964$$

Substituting in Equation A2.10, we obtain,

$$(1 + E_{1,2})^2 = 1.206 \pm 0.484$$

Giving,

$$E_{11} = 0.30 \quad \text{and} \quad E_{22} = -0.15$$

The principal direction *measured from the undeformed axis, Ox*, is, from Equation A2.11,

$$\theta = \frac{1}{2}\tan^{-1}\left(\frac{2 \times 0.419}{1.447 - 0.964}\right) = 30°$$

and from Figure A2.5, this will be measured counterclockwise from *Ox*.

Solution B

This is performed using the properties illustrated in Figure A2.5.

In Figure A2.8, the angle, ψ, in the deformed line, Ob', is,

$$\tan^{-1}(0.348/0.918) = 20.76^\circ$$

As both sets of axes in this diagram are orthogonal,

$$\psi = \gamma_1 + \gamma_2; \quad \text{or,} \quad \gamma_2 = \psi - \gamma_1$$

The lengths of the deformed unit pair are,

$$Oa' = 1.203 \quad Ob' = \sqrt{u_b^2 + v_b^2} = 0.982$$

Using the property that the Cartesian components of shear strain are equal,

$$E_{xy} = Oa' \sin\gamma_1 = Ob' \sin\gamma_2 = Ob' \sin(\psi - \gamma_1)$$
$$= Ob'(\sin\psi\cos\gamma_1 - \cos\psi\sin\gamma_1)$$

Evaluating the above, we obtain,

$$\gamma_1 = \tan^{-1}\frac{0.348}{2.121} = 9.32^\circ \quad \text{and} \quad \gamma_2 = 20.76 - 9.32 = 11.44^\circ$$

From Figure A2.5, we obtain,

$$E_{xy} = Oa' \sin\gamma_1 = 1.203 \sin 9.32 = 0.195$$
$$E_{xx} = Oa' \cos\gamma - 1 = 1.203 \cos 9.32 - 1 = 0.187$$
$$E_{yy} = Ob' \cos\gamma_2 - 1 = 0.982 \cos 11.44 - 1 = -0.0375$$

Substituting in Equation A2.7, we obtain,

$$E_{1,2} = 0.075 \pm 0.225$$

i.e., $E_{11} = 0.30$; $\quad E_{22} = -0.15$

From Equation A2.8, the principal direction makes an angle with the axis Ox of,

$$\theta = \frac{1}{2}\tan^{-1}\left(\frac{2 \times 0.195}{0.187 + 0.0375}\right) = 30^\circ.$$

Ex. A2.5.1 (b) From Equations A2.6,

$$E_{xx} = 0.30\cos^2 30 - 0.15\sin^2 30 = 0.187$$
$$E_{yy} = 0.30\sin^2 30 - 0.15\cos^2 30 = -0.0375$$
$$E_{xy} = \frac{0.30 + 0.15}{2}\sin 60 = 0.195$$

Note that these components are also found as part of the geometric solution in Solution B above.

Solutions to exercises

Chapter 1

Ex. 1.1

(a) $$\text{Initial yield stress } (\sigma_f)_0 = \frac{P_y}{A_0} = \frac{2.89\,\text{kN}}{12.5\,\text{mm}^2} = 231.2\,\text{MPa}$$

$$\text{Strain at initial yield } e_y = \frac{\Delta l}{l_0} = \frac{0.0563}{50} = 0.00112$$

$$\text{Elastic modulus } E = \frac{(\sigma_f)_0}{e_y} = \frac{231.2\,\text{MPa}}{0.001\,12} = 205.33\,\text{GPa}$$

(b) R-value

$$R = \frac{\ln \dfrac{w}{w_0}}{\ln \dfrac{w_0 l_0}{w l}} = \frac{\ln \dfrac{11.41}{12.5}}{\ln \dfrac{12.5 \times 50}{11.41 \times 57.5}} = 1.88$$

Ex. 1.2

Instant area: $A = \dfrac{A_0 l_0}{l}$

Stress: $\sigma = \dfrac{P}{A} = \dfrac{P}{A_0} \dfrac{l}{l_0}$

At 4% elongation,

$$\sigma_A = \frac{P_A}{A_0} \frac{l_A}{l_0} = \frac{1.59\,\text{kN}}{14} \times \frac{52}{50} = 118.11\,\text{MPa}$$

$$\varepsilon_A = \ln\left(\frac{l_A}{l_0}\right) = 0.0392$$

At 8% elongation,

$$\sigma_B = \frac{P_B}{A_0}\frac{l_B}{l_0} = \frac{1.66\,\text{kN}}{14}\frac{54}{50} = 128.06\,\text{MPa}$$

$$\varepsilon_B = \ln\left(\frac{l_B}{l_0}\right) = 0.0769$$

Therefore,

$$n = \frac{\ln(\sigma_A) - \ln(\sigma_B)}{\ln(\varepsilon_A) - \ln(\varepsilon_B)} = 0.12$$

$$K = \frac{\sigma_A}{(\varepsilon_A)^n} = 174.21\,\text{MPa}$$

Ex. 1.3

$$\sigma = K\varepsilon^n\dot{\varepsilon}^m$$

$$\varepsilon = ln\left(\frac{l}{l_0}\right) = \ln\left(\frac{55}{50}\right) = 0.0953$$

$$\dot{\varepsilon}_1 = \frac{0.5\times10^{-3}}{60}/55\times10^{-3} = 0.1515\times10^{-3}\,\text{s}^{-1}$$

$$\dot{\varepsilon}_2 = \frac{50\times10^{-3}}{60}/55\times10^{-3} = 15.15\times10^{-3}\,\text{s}^{-1}$$

$$\sigma_1 = 459\,\text{MPa}, \qquad \sigma_2 = 480\,\text{MPa}$$

$$\Delta\sigma = \sigma_2 - \sigma_1 = 21\,\text{MPa}$$

$$\Delta P = \Delta\sigma A_0\frac{l_0}{l} = 0.27\,\text{kN}$$

Ex. 1.4

By plotting the data in the engineering stress–engineering strain, true stress and true strain curves, the following properties are obtained.

Initial yield stress:	156 MPa
Ultimate tensile strength:	294 MPa
True strain at maximum load:	0.24
Total elongation:	45.4%
Strength coefficient, K:	530 MPa
Strain hardening index, n:	0.24

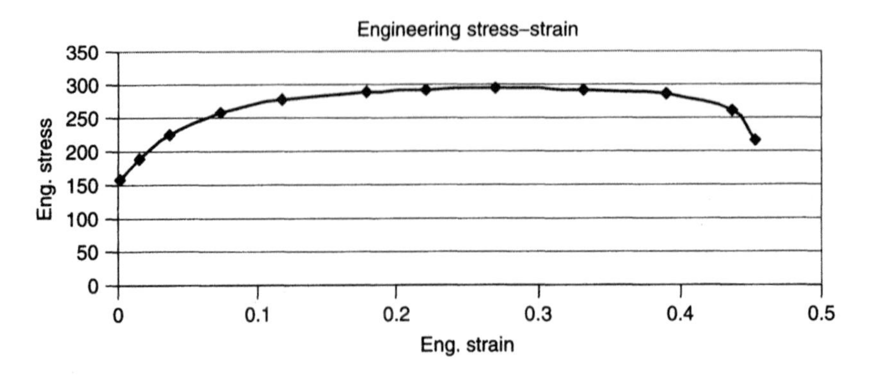

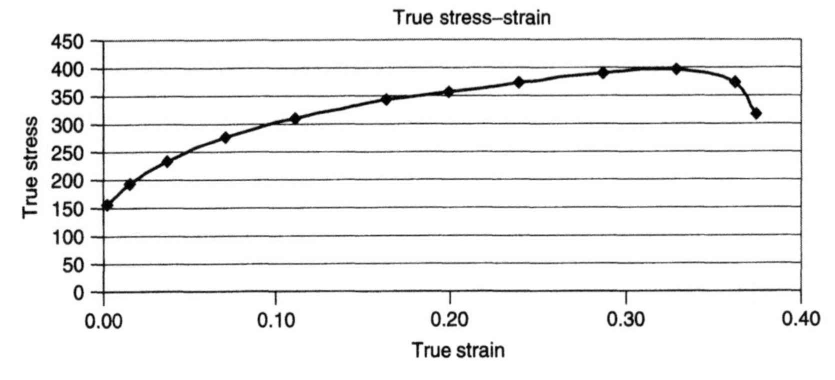

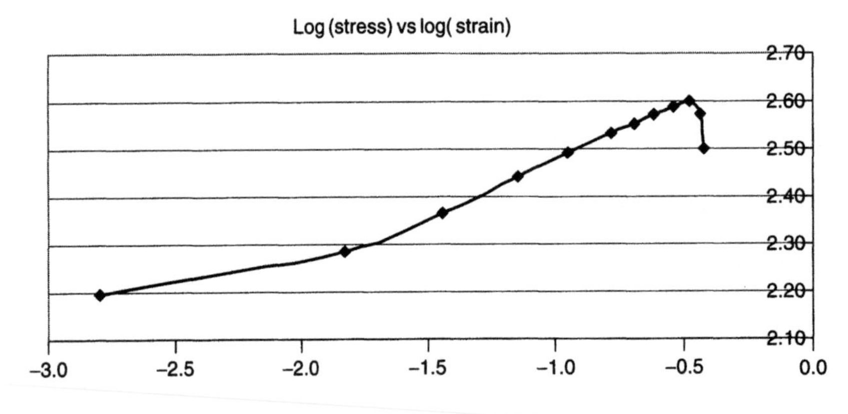

Chapter 2

Ex. 2.1

- Final thickness:

  from incompressibility,

  $lwt = l_0 w_0 t_0 = 6.5 \times 9.4 \times t = 8 \times 8 \times 0.8$

  $t = 0.838\,\text{mm}$

- The principal strains:

$$\varepsilon_1 = \ln\left(\frac{9.4\,\text{mm}}{8\,\text{mm}}\right) = 0.161$$

$$\varepsilon_2 = \ln\left(\frac{6.5\,\text{mm}}{8\,\text{mm}}\right) = -0.208$$

$$\varepsilon_3 = -(\varepsilon_1 + \varepsilon_2) = 0.047$$

Strain ratio:

$$\beta = \frac{\varepsilon_2}{\varepsilon_1} = \frac{-0.208}{0.161} = -1.29$$

Stress ratio:

$$\alpha = \frac{2\beta + 1}{2 + \beta} = -2.225$$

- Membrane stresses:

effective strain,

$$\bar{\varepsilon} = \sqrt{\frac{4}{3}\,(1 + \beta + \beta^2)}\varepsilon_1 = 0.218$$

effective stress,

$$\bar{\sigma} = 600\,(0.008 + \bar{\varepsilon})^{0.22} = 432.57\,\text{MPa}$$

$$\sigma_1 = \frac{\bar{\sigma}}{\sqrt{1-\alpha+\alpha^2}} = 151.3\,\text{MPa}$$

$$\sigma_2 = \alpha\sigma_1 = -336.6\,\text{MPa}$$

Hydrostatic stress $\sigma_\text{h} = \dfrac{\sigma_1 + \sigma_2 + \sigma_3}{3} = -61.7\,\text{MPa}$

Deviatoric stresses:

$$\sigma_1' = \sigma_1 - \sigma_h = 151.3 - (-61.7) = 213\,\text{MPa}$$

$$\sigma_2' = \sigma_2 - \sigma_h = -274.9\,\text{MPa}$$

$$\sigma_3' = \sigma_3 - \sigma_h = 61.7\,\text{MPa.}$$

Check for flow rule:

$$\frac{\varepsilon_1}{\sigma_1'} = \frac{0.161}{213} = 0.000\,756$$

$$\frac{\varepsilon_2}{\sigma_2'} = \frac{-0.208}{-274.9} = 0.000\,756$$

$$\frac{\varepsilon_3}{\sigma_3'} = \frac{0.047}{61.7} = 0.00\,076$$

- Plastic work of deformation,

$$\frac{w}{\text{vol}} = \int_0^{0.218} 600\,(0.008 + \bar{\varepsilon})^{0.22}\,\mathrm{d}\bar{\varepsilon} = \frac{600}{1.22}\,(0.226^{1.22} - 0.008^{1.22})$$

$$= 78.77 \times 10^6\ \text{J/m}^3$$

$$\text{vol} = 51.2\,\text{mm}^3 = 51.2 \times 10^{-9}\,\text{m}^3$$

$$w = 4.03\,\text{J}$$

Ex. 2.2

Diameter of Mohr circle $(\sigma_1 - \sigma_3)$

Yielding with a von Mises criterion,

$$\sqrt{\frac{(\sigma_1 - \sigma_2)^2 + (\sigma_2 - \sigma_3)^2 + (\sigma_3 - \sigma_1)^2}{2}} = \sigma_f$$

Replacing

$$\sigma_2 = \frac{\sigma_1 + \sigma_3}{2},$$

we get that

$$(\sigma_1 - \sigma_3) = \frac{2}{\sqrt{3}}\sigma_f$$

Ex. 2.3

Effective stress,

$$\bar{\sigma} = \sqrt{\frac{(\sigma_1 - \sigma_2)^2 + (\sigma_2 - \sigma_3)^2 + (\sigma_3 - \sigma_1)^2}{2}} = \sigma_f$$

Therefore, for

$$\sigma_2 = \sigma_3 = 0$$

$$\bar{\sigma} = \sigma_1 = \sigma_f$$

From flow rule,

$$\frac{\varepsilon_1}{\sigma_1'} = \frac{\varepsilon_2}{\sigma_2'} = \frac{\varepsilon_3}{\sigma_3'}; \qquad \sigma_1' = \frac{2}{3}\sigma_1; \qquad \sigma_2' = -\frac{\sigma_1}{3}; \qquad \sigma_3' = -\frac{\sigma_1}{3}$$

Therefore,

$$\varepsilon_2 = \varepsilon_3 = -\frac{\varepsilon_1}{2}$$

Effective strain,

$$\bar{\varepsilon} = \sqrt{\frac{2}{3}\left(\varepsilon_1^2 + \varepsilon_2^2 + \varepsilon_3^2\right)} = \varepsilon_1$$

Ex. 2.4

(a) Stress and strain ratios

$$\sigma_1 = 400\,\text{MPa}, \qquad \sigma_2 = 200\,\text{MPa}$$

$$\therefore \alpha = \tfrac{1}{2} \quad \text{and} \quad \beta = 0$$

Therefore,

$$\varepsilon_2 = 0$$

$$\varepsilon_3 = -\varepsilon_1 \quad \Rightarrow \quad \frac{\varepsilon_1}{\varepsilon_3} = -1$$

(b) Under fluid pressure, $-250\,\text{MPa}$.

$\dfrac{\varepsilon_1}{\varepsilon_3} = -1$ will not change.

This can be proven from the flow rule,

$$\sigma_1 = 400 - 250 = 150\,\text{MPa}, \quad \sigma_2 = 200 - 250 = -50\,\text{MPa}$$

$$\sigma_3 = 0 - 250 = -250\,\text{MPa}$$

Mean stress and deviatoric stresses:

$$\sigma_h = \frac{150 - 50 - 250}{3} = -50\,\text{MPa},$$

$$\sigma_1' = 150 - (-50) = 200\,\text{MPa}, \sigma_3' = -250 - (-50) = -200\,\text{MPa}$$

$$\frac{\varepsilon_1}{\sigma_1'} = \frac{\varepsilon_3}{\sigma_3'}$$

therefore,

$$\frac{\varepsilon_1}{\varepsilon_3} = -1$$

This is because yielding is independent of the average stress on an element. Therefore, the strains are also independent.

Ex. 2.5

	$\beta = 1$	$\beta = -1$
Major strain, $\varepsilon_1 = \ln(12/10) =$	0.182	0.182
Minor strain, $\varepsilon_2 = \beta.\varepsilon_1 =$	0.182	−0.182

	$\beta = 1$	$\beta = -1$
Thickness strain, $\varepsilon_3 = -(1+\beta)\varepsilon_1$	−0.365	0
Effective strain, $\bar{\varepsilon} = \sqrt{(4/3)(1+\beta+\beta^2)}\varepsilon_1$	0.365	0.211
Effective stress, $\bar{\sigma} = 850\bar{\varepsilon}^{0.16}$	723	662 MPa
Strain ratio, $\alpha = (2\beta+1)/(2+\beta)$	1	−1
Major stress, $\sigma_1 = \bar{\sigma}/\sqrt{1-\alpha+\alpha^2}$	723	382 MPa
Minor stress, $\sigma_1\alpha$	723	−382 Mpa
Thickness, $t = t \exp \varepsilon_3$	0.833	1.2 mm

Ex. 2.5

Strain/ Stress state	σ_1	σ_2	σ_3	ε_1	ε_2	ε_3
Plane stress pure shear	$\frac{\sigma_f}{\sqrt{3}}$	$-\frac{\sigma_f}{\sqrt{3}}$	0	ε_1	$-\varepsilon_1$	0
Plane stress plane train	$\frac{2}{\sqrt{3}}\sigma_f$	$\frac{1}{\sqrt{3}}\sigma_f$	0	ε_1	0	$-\varepsilon_1$
Plane stress biaxial tension	σ_f	σ_f	0	ε_1	ε_1	$-2\varepsilon_1$

Chapter 3

Ex. 3.1

(c) Principal strains

$$\varepsilon_1 = \ln\left(\frac{6.1\,\text{mm}}{5.0\,\text{mm}}\right) = 0.199, \qquad \varepsilon_2 = \ln\left(\frac{4.8\,\text{mm}}{5.0\,\text{mm}}\right) = -0.041$$

$$\varepsilon_3 = -(\varepsilon_1 + \varepsilon_2) = -0.158$$

$$\beta = \frac{\varepsilon_2}{\varepsilon_1} = -0.21 \Rightarrow \quad \alpha = \frac{2\beta+1}{2+\beta} = 0.324 \text{ (ratio of stress)}$$

Effective strain

$$\bar{\varepsilon} = \sqrt{\frac{4}{3}\left(1+\beta+\beta^2\right)}\varepsilon_1 = 0.21$$

Tensions

$$T_1 = \sigma_1 t = \frac{K \cdot \bar{\varepsilon}^n}{\sqrt{1-\alpha+\alpha^2}} t_0 \exp\left(-(1+\beta)\,\varepsilon_1\right)$$

$$= \frac{600 \times 0.21^{0.22}}{\sqrt{1-0.324+0.324^2}} \times 0.8 \times \exp(-(1-0.21) \times 0.199) = 329.3\text{(kN/m)}$$

$$T_2 = \alpha T_1 = 0.324 \times 248.6 = 108.2\text{(kN/m)}$$

Ex. 3.2

(a) Constant thickness

$$\beta = -1 \quad \Rightarrow \quad \boxed{\alpha = -1}$$

$$\bar{\sigma} = \sqrt{(1-\alpha+\alpha^2)}\,\sigma_1 = \sqrt{3}\sigma_1$$

$$\tau_{max} = \frac{\sigma_1 - \sigma_2}{2} = \sigma_1$$

Therefore

$$\frac{\bar{\sigma}}{\tau_{max}} = \sqrt{3}$$

(b) Uniaxial tension

$$\beta = -\tfrac{1}{2} \quad \Rightarrow \quad \boxed{\alpha = 0}$$

$$\bar{\sigma} = \sqrt{(1-\alpha+\alpha^2)}\,\sigma_1 = \sigma_1$$

$$\tau_{max} = \frac{\sigma_1 - \sigma_2}{2} = \frac{\sigma_1}{2}$$

Therefore

$$\frac{\bar{\sigma}}{\tau_{max}} = 2$$

(c) Plane strain

$$\beta = 0 \quad \Rightarrow \quad \boxed{\alpha = \frac{1}{2}}$$

$$\bar{\sigma} = \sqrt{(1-\alpha+\alpha^2)}\,\sigma_1 = \frac{\sqrt{3}}{2}\sigma_1$$

$$\tau_{max} = \frac{\sigma_1 - \sigma_3}{2} = \frac{\sigma_1}{2}$$

Therefore,

$$\frac{\bar{\sigma}}{\tau_{max}} = \sqrt{3}$$

Ex. 3.3

Location	(a)	(b)	(c)
Strains:	0.015, 0.015	0.050, 0.000	0.150, −0.100
Strain ratio β:	1	0	−0.667
Stress ratio $\alpha = \dfrac{2\beta+1}{2+\beta}$:	1	1/2	−0.251

Thickness strain $-(1+\beta)\varepsilon_1$:	−0.030	−0.050	−0.050	
Thickness: $t = t_0 \exp(\varepsilon_3)$:	0.485	0.476	0.476	(mm)
Stress $\sigma_1 = \dfrac{300}{\sqrt{1-\alpha+\alpha^2}}$:	300	346	262	(MPa)
Tension:	146	165	125	(kN/m)

Ex. 3.4

$$\frac{WD}{\text{Vol}} = \int_0^{\bar{\varepsilon}} \bar{\sigma}\,\mathrm{d}\bar{\varepsilon} = \int_0^{\bar{\varepsilon}} K\,(\varepsilon_0 + \bar{\varepsilon})\,\mathrm{d}\bar{\varepsilon}$$

$$= \frac{K}{1+n}\,(\varepsilon_0+\bar{\varepsilon})^{n+1}\Big|_0^{\bar{\varepsilon}} = \frac{K}{1+n}\left\{(\varepsilon_0+\bar{\varepsilon})^{n+1} - \varepsilon_0^{n+1}\right\}$$

$$= \frac{600\times 10^6}{1.22}\left\{0.04^{1.22} - 0.01^{1.22}\right\} = 7.90\times 10^6 (\mathrm{J/m^3})$$

Work done in unit mass:

$$\frac{WD}{kg} = \frac{7.9\times 10^6}{7850} = 1.01\times 10^3 (\mathrm{J})$$

Temperature increase:

$$\Delta T C_P = J$$

$$\Delta T = \frac{J}{C_P} = \frac{1.01\times 10^3}{0.454\times 10^3} \approx 2.2^\circ\mathrm{C}$$

Chapter 4

Ex. 4.1

$$T_1 = \frac{2Kt_0}{\sqrt{3}}\left(\frac{2}{\sqrt{3}}\varepsilon_1\right)^n \exp(-\varepsilon_1)$$

$$\left(\frac{2}{\sqrt{3}}\varepsilon_1\right)^n \exp(-\varepsilon_1) = \frac{T_1\sqrt{3}}{2Kt_0} = \frac{\sqrt{3}\times 340\times 10^3}{2\times 700\times 10^6\times 0.8\times 10^{-3}} = 0.5258$$

By trial and error,

ε_1	0.08	0.07	0.06
	0.547	0.536	0.523

Therefore,

$\varepsilon_1 = 0.06$

Ex. 4.2

Maximum wall tension is when $\varepsilon_1 = n$.

$$T_{1\,\max} = \frac{2Kt_0}{\sqrt{3}}\left(\frac{2}{\sqrt{3}}n\right)^n \exp(-n)$$

Tension at centre and at binder:

$$T_{1,\mathrm{O}} = T_{1,\mathrm{B}} = \frac{T_{1,\max}}{\exp(\mu\frac{\pi}{2})} = \frac{T_{1,\max}}{\exp\ (0.15\frac{\pi}{2})}$$

$$= \frac{2Kt_0}{1.266 \times \sqrt{3}}\left(\frac{2}{\sqrt{3}}0.2\right)^{0.2} \exp(-0.2) = \frac{2Kt_0}{\sqrt{3}}\left(\frac{2}{\sqrt{3}}\varepsilon_1\right)^{0.2} \exp(-\varepsilon_1)$$

$$\therefore \quad (\varepsilon_1)^{0.2} \exp(-\varepsilon_1) = \frac{1}{1.266}(0.2)^{0.2} \exp(-0.2) = 0.469$$

ε_1 :	0.02	0.03	0.04	0.025	0.026
	0.448	0.481	0.505	0.466	0.469

$\varepsilon_1 = 0.026$

$$T_{1,\mathrm{O}} = T_{1,\mathrm{B}} = \frac{2 \times 600 \times 0.8 \times 10^3}{\sqrt{3}}\left(\frac{2}{\sqrt{3}}0.026\right)^{0.2} \exp(-0.026) = 267(\mathrm{kN})$$

$$B = \frac{267}{2 \times 0.15} = 890\,\mathrm{kN}, \quad 2B = 1780\,\mathrm{kN}$$

Ex. 4.3

Side-wall tension: $1.266T_{1,\mathrm{O}} = 338(\mathrm{kN})$.

Punch force: 676(kN)

$$\frac{\text{Blank-holder force}}{\text{Punch force}} = \frac{1780}{676} = 2.6$$

Ex. 4.4

Angle of wrap from O to A is θ where:

$$\sin\theta = \frac{600 - 10}{2000 - 10}, \quad \theta = 17.25^\circ = 0.301(\mathrm{rad})$$

Tension at mid-point,

$$T_{1,\mathrm{O}} = \frac{2 \times 400 \times 0.8 \times 10^3}{\sqrt{3}}\left(\frac{2}{\sqrt{3}}0.025\right)^{0.17} \exp(-0.025) = 197(\mathrm{kN/m})$$

Tension at A,

$$T_{1,\mathrm{A}} = T_{1,\mathrm{O}} \times \mathrm{e}^{0.1\times0.301} = 197 \times 1.03 = 203(\mathrm{kN/m})$$

At B, angle of wrap is 60 degrees (1.047 rad),

$$T_{1,B} = T_{1,O} \times e^{0.1\times1.047} = 197 \times 1.11 = 219(\text{kN/m})$$

Chapter 5

Ex. 5.1

For the tensile strip, the load is

$$P = \sigma_1 A_1 = K(\varepsilon_0 + \varepsilon)^n A_0 \frac{L_0}{L_1} = K(\varepsilon_0 + \varepsilon)^n t_0 w_0 \exp(-\varepsilon)$$

$$\frac{1}{K t_0 w_0}\frac{\partial P}{\partial \varepsilon} = \frac{\partial}{\partial \varepsilon}\left[(\varepsilon_0 + \varepsilon)^n \exp(-\varepsilon)\right]$$

$$= n(\varepsilon_0 + \varepsilon)^{n-1}\exp(-\varepsilon) - (\varepsilon_0 + \varepsilon)^n \exp(-\varepsilon) = 0$$

$$n = \varepsilon_0 + \varepsilon$$

$$\varepsilon = n - \varepsilon_0$$

Ex. 5.2

For the tensile strip, the load is

$$P = \sigma_1 A_1 = K\varepsilon_1^n A_0 \frac{L_0}{L_1} = K\varepsilon_1^n t_0 w_0 \exp(-\varepsilon_1)$$

For the narrow section, at maximum load $\varepsilon = n$, and

$$P_{\text{max}} = K t_0 w_0 n^n \exp(-n_1)$$

$$= 750 \times 10^6 \times 1.2 \times 12.4 \times 10^{-6}(0.22)^{-0.22}\exp(-0.22) = 6.42(\text{kN})$$

For the wider section, the strain in A is ε_{1A}.

$$P = K\varepsilon_{1A}^n t_0 w_0 \exp(-\varepsilon_{1A})$$

at maximum load,

$$P = K t_0 w_0 \varepsilon_{1A}^n \exp(-\varepsilon_{1A})$$

$$= 750 \times 10^6 \times 1.2 \times 12.5 \times 10^{-6}(\varepsilon_{1A})^{-.22}\exp(-\varepsilon_{1A}) = 6.42(\text{kN})$$

Therefore,

$$f(\varepsilon_{1A}) = \varepsilon_{1A}^{0.22}\exp(-\varepsilon_{1A}) = \frac{12.4}{12.5}0.22^{0.22}\exp(-0.22) = 0.571$$

By iteration,

ε_{1A}	$f(\varepsilon_{1A})$
0.2	0.5746
0.18	0.5728
0.16	0.5694
0.17	0.571

Therefore, $\varepsilon_{1A} = 0.17$, and the 20 mm length becomes:

$l = 20\exp(\varepsilon_{1A}) = 23.7(\text{mm})$

Ex. 5.3

From Equation 5.13, $\dfrac{dA_0}{A_0} = \dfrac{-0.2}{10} = -0.02$

$$(n - \varepsilon_U) \approx \sqrt{-n\frac{dA_o}{A_0}} = \sqrt{0.02n}$$

$$(n - \varepsilon_u)^2 = 0.02n$$

$$n^2 - 2n\varepsilon_u + \varepsilon_u^2 = 0.02n$$

$$n^2 - (2\varepsilon_u + 0.02)n + \varepsilon_u^2 = 0$$

$$n = \frac{(2\varepsilon_u + 0.02) \pm \sqrt{(2\varepsilon_u + 0.02)^2 - 4\varepsilon_u^2}}{2}$$

$$= \frac{(2\varepsilon_u + 0.02) \pm \sqrt{0.08\varepsilon_u + 0.0004}}{2}$$

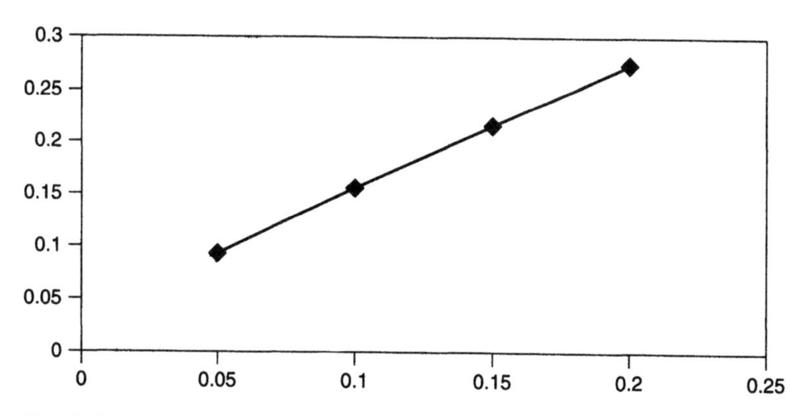

Ex. 5.4

As the load transmitted is equal in each section,

$$P = \sigma_a A_a = \sigma_b A_b$$

$$\sigma_a = \frac{A_b}{A_a}\sigma_b$$

But, $\sigma = B\dot{\varepsilon}^m$

$$B\dot{\varepsilon}_{\mathrm{a}}^m = \frac{A_{\mathrm{b}}}{A_{\mathrm{a}}} B\dot{\varepsilon}_b^m$$

$$\dot{\varepsilon}_{\mathrm{a}} = \left(\frac{A_{\mathrm{b}}}{A_{\mathrm{a}}}\right)^{1/m} \dot{\varepsilon}_{\mathrm{b}}, \text{ or } \dot{\varepsilon}_{\mathrm{b}} = \left(\frac{A_{\mathrm{a}}}{A_{\mathrm{b}}}\right)^{1/m} \dot{\varepsilon}_{\mathrm{a}}$$

In terms of elongation, the total elongation is δL, where

$$\delta L = \delta l_{\mathrm{a}} + \delta l_{\mathrm{b}} = l\,(\dot{\varepsilon}_{1\mathrm{a}} + \dot{\varepsilon}_{1\mathrm{b}})\,\delta t$$

$$v = \frac{\delta L}{\delta t} = l\,(\dot{\varepsilon}_{1\mathrm{a}} + \dot{\varepsilon}_{1\mathrm{b}}) = l\,(\dot{\varepsilon}_{1\mathrm{a}} + \dot{\varepsilon}_{1\mathrm{b}}) = l\dot{\varepsilon}_{1\mathrm{a}}\left(1 + \left(\frac{A_{\mathrm{a}}}{A_{\mathrm{b}}}\right)^{1/m}\right).$$

Therefore,

$$\dot{\varepsilon}_{1\mathrm{a}} = \frac{v}{l\left[1 + (A_{\mathrm{a}}/A_{\mathrm{b}})^{1/m}\right]};\ \dot{\varepsilon}_{1\mathrm{b}} = \frac{v}{l\left[1 + (A_{\mathrm{b}}/A_{\mathrm{a}})^{1/m}\right]}$$

Ex. 5.5

For both elements, the material has zero plastic strain so that the effective stress is

$$\overline{\sigma} = 600(0.004)^{0.2} = 198.9\,\text{MPa}$$

$$\sigma_{1\mathrm{A}} = 198.9\,\text{MPa}, \quad \sigma_{1\mathrm{B}} = \frac{\sigma_{1\mathrm{A}}}{f_0} = \frac{198.9}{0.995} = 199.9\,\text{MPa}$$

Strain ratio is α, then from Equation 2.18(b),

$$\overline{\sigma} = \left(\sqrt{1 - \alpha + \alpha^2}\right)\sigma_1$$

$$\frac{\overline{\sigma}}{\sqrt{1 - \alpha + \alpha^2}} = \sigma_{1\mathrm{B}} = \frac{\overline{\sigma}}{f_0}$$

$$\therefore \sqrt{1 - \alpha + \alpha^2} = f_0 = 0.995, \quad \alpha = 0.99$$

$$\sigma_{2\mathrm{B}} = \alpha\sigma_{1\mathrm{B}} = 0.99 \times 199.9 = 197.9(\text{MPa})$$

Chapter 6

Ex. 6.1

(a) Determine the limiting elastic curvature

$$\frac{M}{I} = E'\left(\frac{1}{\rho}\right) = \frac{\sigma_1}{y}$$

where

$$I = \frac{wt^3}{12}$$

At limit of elastic bending, $\sigma_1 = \sigma_f$

Limiting elastic curvature:

$$M_e = I\frac{S}{y} = S\frac{wt^2}{6} = \frac{250 \times 10^6 \times 50 \times 10^{-3} \times (2 \times 10^{-3})^2}{6} = 8.33(\text{Nm})$$

$$\left(\frac{1}{\rho}\right) = \frac{M_e}{EI} = \frac{8.33}{200 \times 10^9 \times \dfrac{50 \times 10^{-3} \times (2 \times 10^{-3})^3}{12}} = 1.25\,\text{m}^{-1}$$

$$\Rightarrow \quad R = 800\,\text{mm}$$

(b) Construct a moment curvature diagram

Elastic moment: $M_e = EI\left(\frac{1}{\rho}\right)$

Plastic moment: $M_p = S\frac{wt^2}{4} = 12.5\,\text{Nm}$

Elastic plastic moment,

$$\left(\frac{1}{\rho}\right) > \left(\frac{1}{\rho_e}\right)$$

but before the moment reaches fully plastic moment M_p.

From Figure 6.11, the material is in plastic deformation above y_e, and elastic below y_e.

$$\text{Strain} = \frac{S}{E} = y_e\left(\frac{1}{\rho}\right) \quad \Rightarrow \quad y_e = \frac{S}{E\,(1/\rho)}$$

Equilibrium equation in bending,

$$M = 2\int_0^{t/2} \sigma w y\,dy = 2w\left(E\left(\frac{1}{\rho}\right)\int_0^{y_e} y^2\,dy + S\int_{z_1}^{\frac{t}{2}} y dy\right)$$

$$M = 2w\left(E\frac{1}{\rho}\left(\frac{y^3}{3}\right)_0^{y_e} + S\left(\frac{y^2}{2}\right)_{y_e}^{\frac{t}{2}}\right) = w\frac{St^2}{12}(3 - m^2)$$

where $\frac{1}{\rho} = \frac{1}{\rho_e} \cdot \frac{1}{m}$ (see Equation 6.23)

(c) • Using the approximate relation

$$\Delta\left(\frac{1}{\rho}\right) = -3\frac{S}{Et}$$

$$\rho = 400\,\text{mm} \quad \Rightarrow \quad \left(\frac{1}{\rho}\right) = 2.5\,\text{m}^{-1}$$

$$\Delta\left(\frac{1}{\rho}\right) = -3\frac{S}{Et} = 1.88\,\text{m}^{-1}$$

Final curvature

$$\frac{1}{\rho} = 2.5 - 1.88 = 0.625\,\text{m}^{-1}$$

$$\boxed{\rho = 1.6\,\text{m}}$$

- Using the diagram in (b)

$$y_e = \frac{\sigma_f}{E\,(1/\rho)} = 0.5 \times 10^{-3}$$

$$M = 2w\left(E\frac{1}{\rho}\left(\frac{y^3}{3}\right)_0^{y_e} + \sigma_f\left(\frac{y^2}{2}\right)_{y_e}^{\frac{t}{2}}\right) = 11.46\,\text{Nm}$$

$$\Delta\left(\frac{1}{\rho}\right) = \frac{\Delta M}{EI} = 1.72\,\text{m}^{-1}$$

$$\boxed{\rho = 1.28\,\text{m}}$$

Ex. 6.2

Initial radius of curvature and wrap:

$$\rho_0 = \frac{l}{2\pi} \text{ and } \theta_0 = l\frac{1}{\rho_0} = 2\pi$$

Springback due to unloading:

$$\frac{\Delta\left(\frac{1}{\rho}\right)}{\frac{1}{\rho_0}} = \frac{\Delta\theta}{\theta_0} = -\frac{3S}{E't}\rho_0$$

The gap between the ends of the strip after unloading is given by

$$\text{gap} = \rho_0\Delta\theta = \frac{l}{2\pi}\left(-\frac{3S}{E't}\rho_0\theta_0\right) = -\frac{l}{2\pi}\left(\frac{3S}{E't}\right)\frac{l}{2\pi}\cdot 2\pi$$

$$\text{gap} = \frac{3}{2\pi}\frac{l^2}{t}\frac{S}{E'}$$

Ex. 6.3

$$E' = \frac{E}{1-\nu^2} = \frac{200}{1-0.3^2} = 219.8\,\text{GPa}$$

$$\sigma\frac{2}{\sqrt{3}}\overline{\sigma} = \frac{2}{\sqrt{3}}600\left(\frac{2}{\sqrt{3}}\varepsilon\right)^{0.22} = 715\varepsilon^{0.22}\,\text{MPa}$$

$$I_n = \frac{0.2(2 \times 10^{-3})^{2.22}}{2^{1.22} \times 2.22} = 3.94 \times 10^{-8} \qquad \text{from Equation 6.27}$$

From Equation 6.26

$$M = I_n K' \left(\frac{1}{\rho}\right)^n = \frac{3.94 \times 10^{-8} \times 715 \times 10^6}{0.08^{0.22}} = 49.1\,\text{Nm}$$

For elastic unloading

$$\Delta\left(\frac{1}{\rho}\right) = \frac{\Delta M}{E'I} = \frac{-49.1}{219.8 \times 10^9} \times \frac{12}{0.2(2 \times 10^{-3})^3} = 1.675\,\text{m}^{-1}$$

Final curvature is

$$\frac{1}{0.08} - 1.675 = 10.8\,\text{m}^{-1}$$

and the final radius of curvature is $\frac{1}{10.8} = 92\,\text{mm}$

Ex. 6.4

Springback in an elastic perfectly plastic material:

$$\frac{\Delta\left(\frac{1}{\rho}\right)}{\frac{1}{\rho_0}} = -\frac{3S}{E'}\frac{\rho_0}{t}$$

Springback between aluminum and steel can be compared as follows:

$$B = \frac{\Delta\,(1/\rho)/(1/\rho_0)\,|_{\text{Al}}}{\Delta\,(1/\rho)/(1/\rho_0)\,|_{\text{Steel}}} = \frac{S_{\text{Al}}/E'_{\text{Al}}}{S_{\text{Steel}}/E'_{\text{Steel}}} = \frac{S_{\text{Al}}}{S_{\text{Steel}}} \times \frac{E_{\text{Steel}}}{E_{\text{Al}}}$$

Since the material is work-hardening, an approximation can be made by assuming the average stress to be $S = K\varepsilon^n$ at the respective bend ratio.

Therefore,

$$B = \frac{\Delta\,(1/\rho)/(1/\rho_0)\,|_{\text{Al}}}{\Delta\,(1/\rho)/(1/\rho_0)\,|_{\text{Steel}}} = \frac{205\varepsilon^{0.2}}{530\varepsilon^{0.26}} \times \frac{190}{75} = 0.98\,(\varepsilon)^{-0.06}$$

For $\rho/t = 10$, the strain is $\varepsilon = y/\rho = t/2\rho = 0.05$

$B = 1.173$

For $\rho/t = 5$, the strain is $\varepsilon = y/\rho = t/2\rho = 0.1$

$B = 1.125$

Alternative solutions

The springback for a work-hardening material can be derived as

$$\Delta\left(\frac{1}{\rho}\right)=\left(\frac{6}{2+n}\right)\left(\frac{K'}{E'}\right)\left(\frac{t}{2\rho}\right)^n\left(\frac{1}{t}\right)$$

Therefore, the springback ratio between aluminum and steel is

$$B=\frac{\left(\frac{6}{2+n}\right)\left(\frac{K'}{E'}\right)\left(\frac{t}{2\rho}\right)^n|_{\text{Al}}}{\left(\frac{6}{2+n}\right)\left(\frac{K'}{E'}\right)\left(\frac{t}{2\rho}\right)^n|_{\text{Steel}}}=\frac{2+n_{\text{steel}}}{2+n_{\text{Al}}}\times\frac{K_{Al}}{K_{\text{Steel}}}\times\frac{E_{\text{Steel}}}{E_{\text{Al}}}\times\left(\frac{t}{2\rho}\right)^{0.2-0.26}$$

$$=\frac{2.26}{2.2}\times\frac{205}{530}\times\frac{190}{75}\left(\frac{t}{2\rho}\right)^{0.2-0.26}=1.007\left(\frac{t}{2\rho}\right)^{-0.06}$$

For $\rho/t=10$, $B=1.007\,(0.05)^{-0.06}=1.205$

For $\rho/t=5$, $B=1.007\,(0.1)^{-0.06}=1.156$

Chapter 7

Ex. 7.1

In the hole expansion process the equilibrium equations are

$$T_\phi=\overline{T}\left(1-\frac{r_{\text{i}}}{r}\right)$$

Then for $T_\phi=\dfrac{2\overline{T}}{3}$ at $r=r_0$:

$$T_\phi=\overline{T}\left(1-\frac{r_{\text{i}}}{r_0}\right)=\frac{2\overline{T}}{3}\quad\Rightarrow\quad\left(\frac{r_0}{r_{\text{i}}}\right)=3$$

Ex. 7.2

For shells,

$$\frac{\mathrm{d}T_\theta}{\mathrm{d}r}-\frac{T_\theta-T_\phi}{r}=0$$

For nosing, $T_\theta=-\overline{T}$.

Therefore,

$$\frac{\mathrm{d}T_\theta}{\mathrm{d}r}+\frac{\overline{T}+T_\phi}{r}=0$$

$$\frac{\mathrm{d}T_\theta}{T_\phi+\overline{T}}=-\frac{\mathrm{d}r}{r}$$

Integrating the above equation yields the following:

$$\ln\left(T_\phi + \overline{T}\right)\Big|_0^{T_\phi} = -\ln r\,\Big|_{r_i}^{r}$$

$$\ln\left(T_\phi + \overline{T}\right) - \ln\left(\overline{T}\right) = -\ln\left(\frac{r}{r_i}\right)$$

$$T_\phi = -\overline{T}\left(1 - \frac{r_i}{r}\right)$$

Ex. 7.3

In flaring the equilibrium equations, given the boundary conditions $T_\phi = 0$ at r_0, are

$$T_\phi = -\overline{T}\ln\left(\frac{r_0}{r}\right)$$

$$T_\theta = \overline{T} + T_\phi = \overline{T}\left(1 - \ln\left(\frac{r_0}{r}\right)\right)$$

Therefore, the limit of T_ϕ is given by $-\overline{T}$ at r_i

$$T_\phi = -\overline{T}\ln\left(\frac{r_0}{r_i}\right) = -\overline{T} \quad \Rightarrow \quad \ln\left(\frac{r_0}{r_i}\right) = 1 \Rightarrow \quad \left(\frac{r_0}{r_i}\right) = \mathrm{e}$$

Then r can move between r_i and the maximum $r_0 = r_i\mathrm{e}$.

$$\boxed{r_i < r < r_i\mathrm{e}}$$

Chapter 8

Ex. 8.1

(a) From Equation 8.11

$$h_f = \frac{r_i}{2}\left\{\left(\frac{r_0}{r_i}\right)^2 - 1\right\} = \frac{25}{2}\left\{\left(\frac{50}{25}\right)^2 - 1\right\} = 37.5\,\text{mm}$$

(b) Max punch force

$$F_d = 2\pi r_i t_0\left\{\sigma_f \ln\left(\frac{r_0}{r_i}\right) + \frac{\mu B}{\pi r_0 t_o}\right\}\exp\left(\frac{\mu\pi}{2}\right)$$

$$F_d = 2\pi \times 25 \times 1.2\left\{350\ln\left(\frac{50}{25}\right) + \frac{0.1 \times 30000}{\pi \times 50 \times 1.2}\right\}\exp\left(\frac{0.1 \times \pi}{2}\right) = 57(\text{kN})$$

Ex. 8.2

The maximum punch load will be lower for $n = 0.2$ than for $n = 0.1$. The curve also moves to the right as shown below.

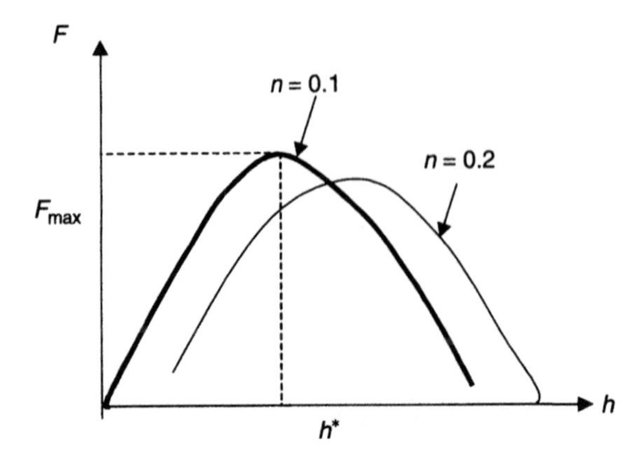

Ex. 8.3

For a small section of the sheet, assume the sheet slides down.

Tool pressure per unit width:

$$p = 2T \sin\left(\frac{d\theta}{2}\right) = T d\theta$$

$$T + dT + \mu P = T$$

$$dT = -\mu T d\theta$$

$$\frac{dT}{T} = -\mu \, d\theta$$

$$\ln\left(\frac{T_2}{T_1}\right) = -\frac{\pi}{2}\mu, \qquad T_2 = T_1 \exp\left(-\frac{\pi}{2}\mu\right)$$

$$T_2 - T_1 = T_1\left[\exp\left(-\frac{\pi}{2}\mu\right) - 1\right]$$

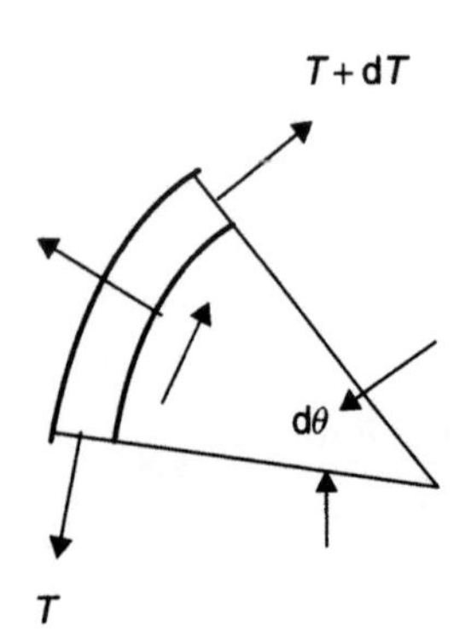

Ex. 8.4

(a) Volume of metal,

$$\pi \frac{d_0^2}{4} t_0 = \pi d_f t_f h_f + \pi \frac{d_f^2}{4} t_0$$

$$d_0 = 120\,\text{mm}$$

$$\text{Total drawn ratio} = \frac{120\,\text{mm}}{62\,\text{mm}} = 1.94$$

Limiting drawn ratio (LDR) $LDR \leq \exp(\eta)$

$$LDR \leq \exp(0.6) = 1.82$$

Therefore, the can cannot be drawn in one operation.

(b) If the draw ratios are equal for the two drawing operations,

$$\frac{d_0}{d_i} = \frac{d_i}{d_f} \quad \Rightarrow \quad d_i = \sqrt{120 \times 62} = 86\,\text{mm}$$

Height from first draw:

$$h_i = \frac{r_i}{2}\left\{\left(\frac{r_0}{r_i}\right)^2 - 1\right\} = 20\,\text{mm}$$

(c) Punch force, neglecting blank holding force and die friction:

$$F_d = \frac{2\pi r_i t_0}{\eta} \sigma_f \ln\left(\frac{d_0}{d_i}\right) = 21.6\,\text{kN}$$

(d) Cup height at the finish of second draw:

$$h_f = \frac{r_f}{2}\left\{\left(\frac{r_0}{r_f}\right)^2 - 1\right\} = 42.6\,\text{mm}$$

(e) The full height is obtained by ironing the middle section.

(f) Saving from reduced thickness:

$$\text{Volume of saving per blank} = \frac{\pi d_0^2}{4}\{t_{1983} - t_{1998}\} = \frac{\pi 120^2}{4}\{0.41 - 0.33\} = 905\,\text{mm}^3$$

For 100 billion blanks, volume saved $= 905 \times 10^2\,\text{m}^3$

Cost of metal saved $= 5 \times 2800 \times 905 \times 10^2 = \$1.25$ billion

Chapter 9

Ex. 9.1

Using the relation in Section 9.1.3, the effective stress is

$$\overline{\sigma} = \frac{6.4}{16}\frac{3}{1.2}\left\{(61/3)^2 + 4\right\}(61/50)^2 = 620\,\text{MPa}$$

and the effective strain is,

$$\overline{\varepsilon} = 2\ln(61/50) = 0.40$$

Ex. 9.2

The principal radius of curvature ρ_1 at and near the tangent point is 50 mm. The principal tensions at the tangent point, from Section 9.2.1, are

$T_\phi = 374\,\text{MPa}$ and $T_\theta = \alpha T_\phi = 0.89 \times 374 = 333\,\text{MPa}$

As indicated in Section 9.2, in the unsupported region, $p = 0$, and therefore,

$$\frac{T_\theta}{\rho_1} = -\frac{T_\phi}{\rho_2}$$

i.e.

$$\rho_2 = -\frac{374}{333} \times 50 = -56\,\text{mm}$$

Ex. 9.3

From Equations 9.4 and 9.6, it may be shown that

$$\frac{t_0}{t} = 1 + \left(\frac{h}{a}\right)^2$$

From Equation 9.1, the effective strain is

$$\overline{\varepsilon} = -\varepsilon_t = \ln(t_0/t) = \ln\left[1 + (h/a)^2\right]$$

and the membrane stress is

$$\sigma_\phi = \sigma_\theta = \overline{\sigma} = 350\overline{\varepsilon}^{0.18}\,\text{MPa}$$

From Equation 9.5, the bulging pressure is

$$p = \frac{2\sigma_\phi t}{\rho} = \frac{2 \times 350\overline{\varepsilon}^{0.18}}{a^2\left[1 + (h/a)^2\right]/2h} \cdot \frac{1.2 \times 10^{-3}}{\left[1 + (h/a)^2\right]}\,\text{MPa}$$

where

$$\rho = \frac{a^2\left[1 + (h/a)^2\right]}{2h}$$

For $a = 50\,\text{mm}$, and computing in the range, $0 < h < 45$, we obtain the characteristic shown below.

(The maximum pressure occurs at a membrane strain of approximately 0.2 ($\bar{\varepsilon} = 0.4$). As indicated in Figure 9.3, for a ductile material this is probably less than the forming limit curve in biaxial tension and it is expected that the sheet would deform beyond this maximum pressure and fail at a higher strain under a falling pressure gradient.)

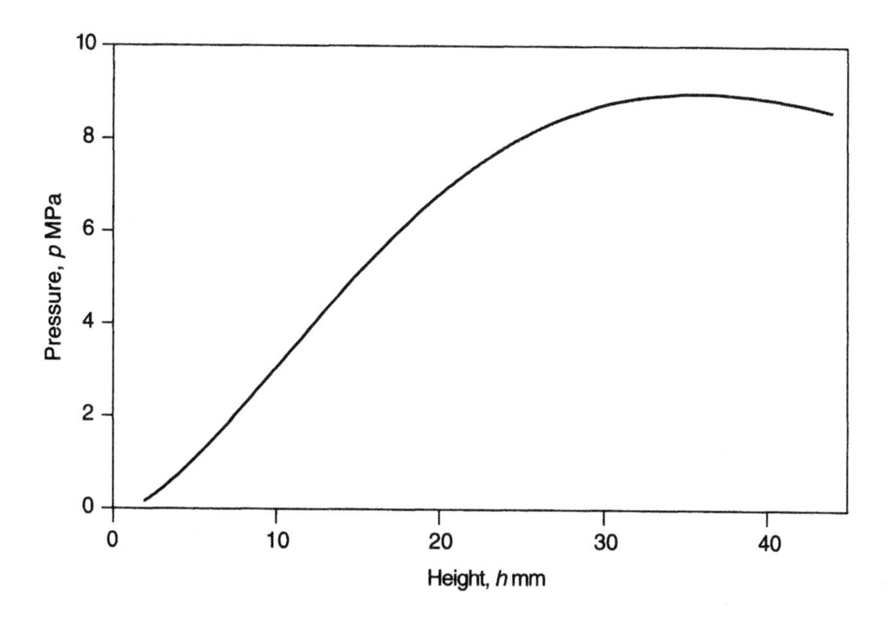

Chapter 10

Ex. 10.1

The stress after bending over the block without tension is, from Equation 6.17,

$$\sigma_1 = \pm\frac{E't}{2\rho_0} = \pm\frac{78 \times 10^9 \times 1.85 \times 10^{-3}}{2 \times 0.6} = 120\,\text{MPa}$$

On the outer (tension) side of the sheet, the additional stress to cause yielding as in Figure 10.2(d), is $180 - 120 = 60\,\text{MPa}$. The tension to be applied to add this stress is

$$T = \Delta\sigma_1 t = 60 \times 10^6 \times 1.85 \times 10^{-3} = 111\,\text{kN/m}$$

The fully plastic tension, as in Figure 10.2(e), is

$$T_y = St = 180 \times 10^6 \times 1.85 \times 10^{-3} = 333\,\text{kN/ m}$$

Ex. 10.2

The limiting elastic radius of curvature is, from Equation 6.19,

$$\rho_e = \frac{E't}{2S} = \frac{78 \times 10^9 \times 1.85 \times 10^{-3}}{2 \times 180 \times 10^6} = 0.401\,\mathrm{m}$$

From Equation 10.2, the moment prior to applying the tension is

$$M_0 = \frac{78 \times 10^9 \left(1.85 \times 10^{-3}\right)^3}{12 \times 0.6} = 68\,\mathrm{Nm/m}$$

From Equation 10.6, when $m = 0$, the moment is

$$M = M_0/2 = 34\,\mathrm{Nm/m}$$

From Equation 10.7, the tension for $m = 0$ is

$$T = T_y\left\{1 - \frac{\rho_e}{4.\rho_0}\right\} = 333\left\{1 - \frac{0.4}{4 \times 0.6}\right\} = 278\,\mathrm{kN/m}$$

As the loading is elastic, the change in curvature on unloading is, from Equation 6.30,

$$\Delta\left(\frac{1}{\rho}\right) = \frac{\Delta M}{E'I} = -\frac{34 \times 12}{78 \times 10^9 \left(1.85 \times 10^{-3}\right)^3} = -0.826$$

The final curvature is

$$\frac{1}{0.6} - 0.826 = 0.84\,\mathrm{m}^1$$

and the final radius of curvature is $1/0.84 = 1.19\,\mathrm{m}$

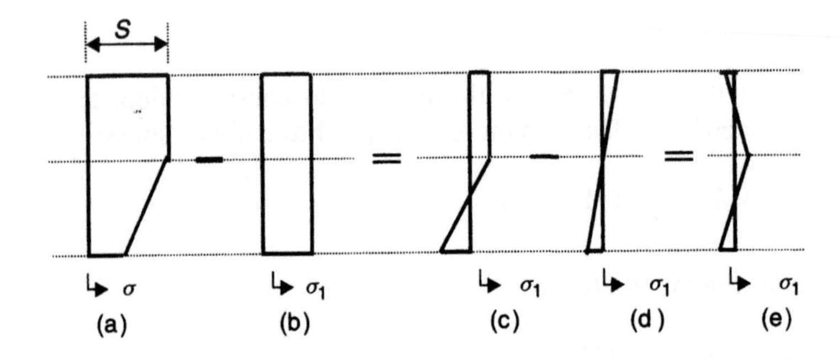

The stress distribution is shown diagrammatically above. The elastic, plastic stress state under the tension and moment, similar to that in Figure 10.2(d), is shown in (a).

Unloading the tension is equivalent to subtracting the uniform stress in (b) to give a distribution in (c) in equilibrium with the moment. Subtracting the elastic bending stress in (d) occurs when the moment is released to give the residual stress in (e).

Ex. 10.3

From Equation 10.12, the slope of the stress–strain curve for steel is

$$\frac{0.2 \times 700 \times 0.012^{0.2}}{0.012} = 4.82 \text{ GPa}$$

and for aluminium, is

$$\frac{0.2 \times 400 \times 0.012^{0.2}}{0.012} = 2.75 \text{ GPa}$$

From Equation 10.11, the springback, i.e. the change in curvature on unloading, is, for steel,

$$\Delta\left(\frac{1}{\rho}\right)_{\text{steel}} = -\frac{4.82}{220}\frac{1}{2.5} = -0.0088\,\text{m}^{-1}$$

and for aluminium,

$$\Delta\left(\frac{1}{\rho}\right)_{\text{al.}} = -\frac{2.75}{78}\frac{1}{2.5} = -0.0141\,\text{m}^{-1}$$

The final curvature for steel is $(1/2.5) - 0.0088 = 0.391$ giving a radius of curvature of 2.56 m. For aluminium, the final curvature is $(1/2.5) - 0.0141 = 0.386$, giving a radius of curvature of 2.59. Clearly the springback is very small and similar for both metals. The result is independent of thickness and depends on the ratio of the slope of the stress–strain curve to the elastic modulus, which are not very different for these two cases.

Ex. 10.4

From Equation 10.20, the tension after the first bend is

$$T = T_0 + \Delta T = 0.6St_0 + \frac{St_0^2}{4bt_0}\left(1 + 0.6^2\right) = St_0\left\{0.6 + \frac{0.34}{b}\right\}$$

The thickness reduction is, from Equation 10.21,

$$\frac{\Delta t}{t_0} = -\frac{t_0}{2bt_0}\left(\frac{T}{T_y}\right) = -\frac{0.6}{2b}$$

The thickness downstream of the first bend is

$$t = t_0 + \Delta t = t_0\left\{1 - \frac{0.3}{b}\right\}$$

The stress in the sheet downstream of the first bend is

$$\frac{\sigma_1}{S} = \frac{T}{St}$$

Evaluating in the range given, we obtain the diagrams below.

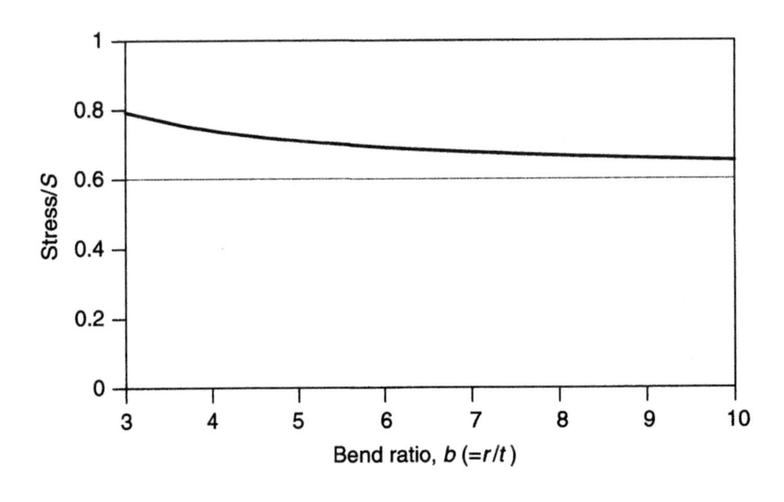

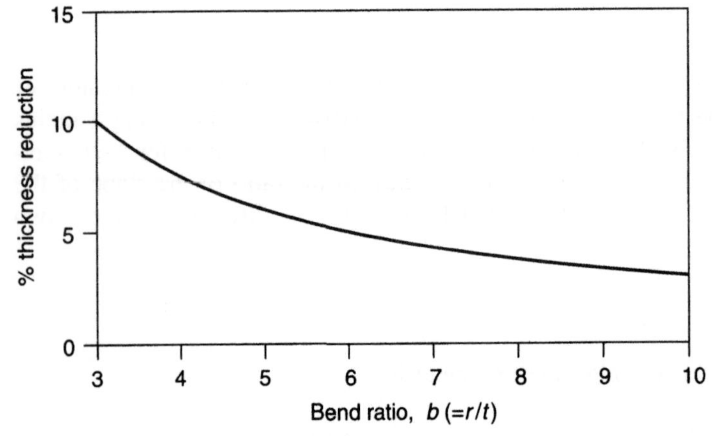

Chapter 11

Ex. 11.1

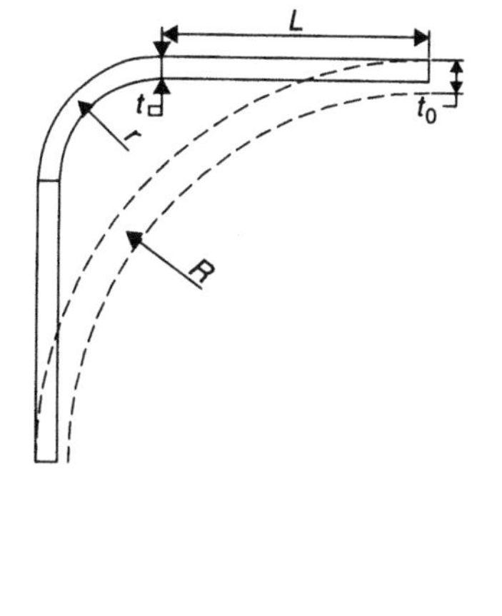

(a) For the frictionless case

The current radius of the tube at the corner is r. The tube wall thickness will be uniform. From volume constancy

$$t\left(2(R-r)+\frac{\pi}{2}r\right)=\frac{\pi}{2}Rt_0$$

re-arranging

$$t=\frac{t_0}{\dfrac{4}{\pi}-\left(\dfrac{4-\pi}{\pi}\right)\dfrac{r}{R}}$$

For the plane strain condition

$$\bar{\varepsilon}=\frac{2}{\sqrt{3}}\varepsilon_\theta=\frac{2}{\sqrt{3}}\ln\left(\frac{t_0}{t}\right)\text{ and }\bar{\sigma}=\frac{\sqrt{3}}{2}\sigma_l$$

$$\sigma_\theta=\frac{2}{\sqrt{3}}K\left[\varepsilon_o+\frac{2}{\sqrt{3}}\ln\left(\frac{4}{\pi}-\left(\frac{4-\pi}{\pi}\right)\frac{r}{R}\right)\right]^n$$

Internal pressure is given by

$$p=\frac{\sigma_\theta t}{r}=\frac{2t_oK}{\sqrt{3}r\left[\dfrac{4}{\pi}-\left(\dfrac{4-\pi}{\pi}\right)\dfrac{r}{R}\right]}\left[\varepsilon_o+\frac{2}{\sqrt{3}}\ln\left(\frac{4}{\pi}-\left(\frac{4-\pi}{\pi}\right)\frac{r}{R}\right)\right]^n$$

(b) For the full sticking case

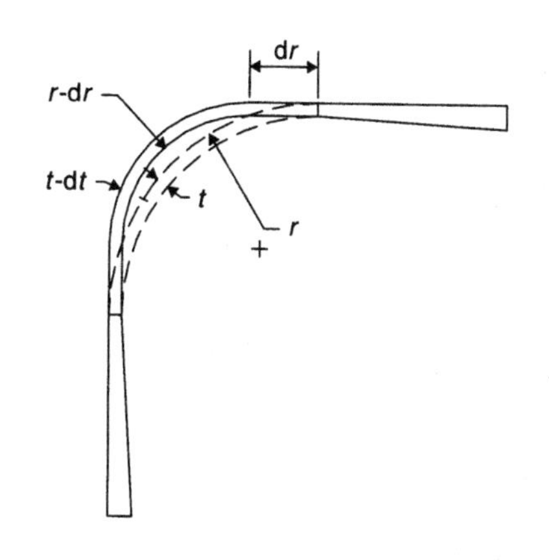

The relation between the corner radius and the corner thickness can be obtained from the incremental relation

$$t\,\mathrm{d}r+\frac{\pi}{2}(r-\mathrm{d}r)(t-\mathrm{d}t)=\frac{\pi}{2}rt$$

$$\int_R^r\left(\frac{4}{\pi}-1\right)\frac{\mathrm{d}r}{r}=\int_{t_o}^t\frac{\mathrm{d}t}{t}$$

$$t=t_o\left(\frac{r}{R}\right)^{\left(\frac{4}{\pi}-1\right)}$$

Circumferential stress

$$\sigma_1 = \frac{2}{\sqrt{3}} K \left[\varepsilon_o + \frac{2}{\sqrt{3}} \left(\frac{4}{\pi} - 1 \right) \ln \left(\frac{R}{r} \right) \right]^n$$

The internal pressure is given by

$$p = \frac{\sigma_1 t}{r} = \frac{2t_0}{\sqrt{3}r} \left(\frac{r}{R} \right)^{\left(\frac{4}{\pi}-1\right)} K \left[\varepsilon_o + \frac{2}{\sqrt{3}} \left(\frac{4}{\pi} - 1 \right) \ln \left(\frac{R}{r} \right) \right]^n$$

Ex. 11.2

The minimum corner radius is limited either by necking or by the maximum fluid pressure available. Since

$$t = \frac{t_0}{\frac{4}{\pi} - \left(\frac{4-\pi}{\pi} \right) \frac{r}{R}}$$

$\varepsilon_\theta = -\varepsilon_t = \ln \frac{t_0}{t}$, $\varepsilon_\theta^* = n = 0.2$ for local necking of tube wall, therefore,

$t = 3.27, \qquad r = 17\,\text{mm}$

The internal pressure required to form corner radius r is given in **Ex. 11.1**. Substituting the material properties and die geometry into the equation, we obtain:

r(mm)	t(mm)	σ_θ(MPa)	p(MPa)
38.00	3.45	566.85	51.52
37.00	3.44	568.86	52.96
36.00	3.44	570.84	54.47
35.00	3.43	572.78	56.08
34.00	3.42	574.70	57.77
33.00	3.41	576.58	59.56
32.00	3.40	578.44	61.46
31.00	**3.39**	**580.27**	**63.48**
30.00	3.38	582.07	65.63
29.00	3.37	583.84	67.92

Therefore, the minimum corner radius is 31 mm, and is limited by the maximum pressure available.

Ex. 11.3

In order to calculate the initial tube thickness, we use volume constancy.

$$2\left(\frac{t_1 + t_2}{2} \right) L + \frac{\pi}{2} r t_2 = \frac{\pi}{2} R t_0$$

where $L = (R - r)$

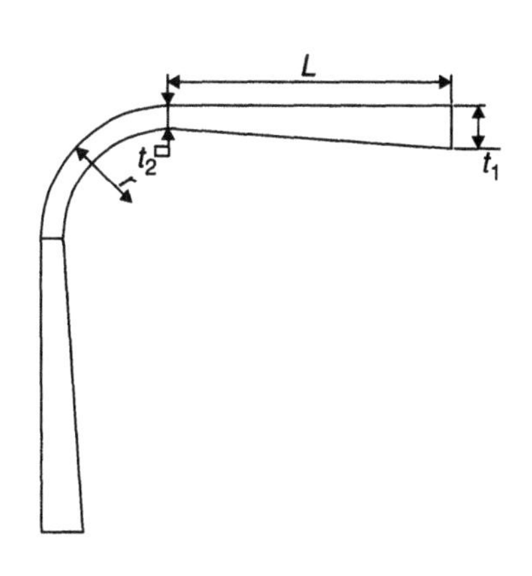

$$t_o = \frac{1}{R}\left[2\left(\frac{t_1 + t_2}{\pi}\right)(R - r) + rt_2\right]$$

The forces that existed at each end of the wall can be calculated as

$$F_1 = \sigma_1 t_1 = \frac{2}{\sqrt{3}} K \left[\varepsilon_o + \frac{2}{\sqrt{3}} \ln\left(\frac{t_0}{t_1}\right)\right]^n t_1$$

$$F_2 = \sigma_2 t_2 = \frac{2}{\sqrt{3}} K \left[\varepsilon_o + \frac{2}{\sqrt{3}} \ln\left(\frac{t_0}{t_2}\right)\right]^n t_2$$

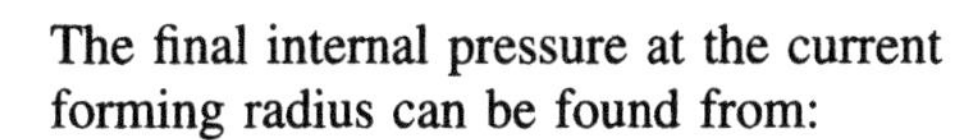

The final internal pressure at the current forming radius can be found from:

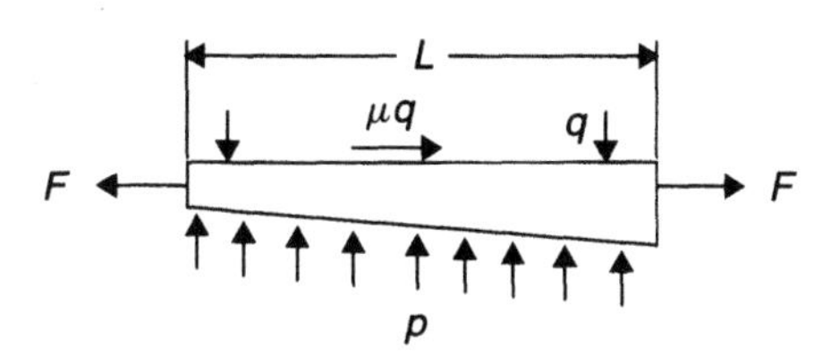

$$F_2 = pr \quad \text{thus} \quad p = \frac{F_2}{r}$$

To calculate the average friction coefficient from force balance

$$\mu_{av} pL = F_2 - F_1 \quad \text{where } q = p$$

$$\mu_{av} = \frac{F_2 - F_1}{pL}$$

Ex. 11.4

(a) For the free ends, there is no stress in the axial direction, $\sigma_2 = 0$.

For a thin cylinder, $\sigma_3 = 0$.

$$\varepsilon_1 = \ln\left(\frac{r}{r_0}\right)$$

$$\varepsilon_2 = \varepsilon_3 = -\tfrac{1}{2}\varepsilon_1$$

$$\ln\left(\frac{t}{t_0}\right) = -\tfrac{1}{2}\ln\left(\frac{r}{r_0}\right)$$

$$\frac{t}{t_0} = \sqrt{\frac{r_0}{r}}$$

Effective stress and strains are

$\overline{\sigma} = \sigma_1$ and $\overline{\varepsilon} = \varepsilon_1$

The internal pressure is calculated as

$$p = \frac{\sigma_1 t}{r}$$

$$p = K t_0 \sqrt{\frac{r_0}{r^{3/2}}} \left[\ln\left(\frac{r}{r_0}\right) \right]^n$$

To calculate the expanded radius at instability, let $\mathrm{d}p = 0$.

This leads to

$$\ln\left(\frac{r}{r_0}\right) = \frac{2n}{3} \text{ or } r = r_0 \mathrm{e}^{2n/3}$$

(b) For restricted ends, there is no strain in the axial direction, $\varepsilon_2 = 0$.

$$\varepsilon_1 = \ln\left(\frac{r}{r_0}\right)$$

$$\varepsilon_3 = -\varepsilon_1$$

$$\ln\left(\frac{t}{t_0}\right) = -\ln\left(\frac{r}{r_0}\right)$$

$$\frac{t}{t_0} = \frac{r_0}{r}$$

Effective stress and strains are:

$$\overline{\sigma} = \tfrac{\sqrt{3}}{2}\sigma_1 \text{ and } \overline{\varepsilon} = \tfrac{2}{\sqrt{3}}\varepsilon_1$$

$$\sigma_1 = \frac{2}{\sqrt{3}} K \left[\frac{2}{\sqrt{3}} \ln\left(\frac{r}{r_0}\right) \right]^n$$

The internal pressure is calculated as

$$p = \frac{\sigma_1 t}{r}$$

$$p = \left(\frac{2}{\sqrt{3}}\right)^{n+1} K \frac{t_0 r_0}{r^2} \left[\ln\left(\frac{r}{r_0}\right) \right]^n$$

To calculate the expanded radius at instability, let $\mathrm{d}p = 0$.

This leads to

$$\ln\left(\frac{r}{r_0}\right) = \frac{n}{2} \text{ or } r = r_0 \mathrm{e}^{n/2}$$

Index